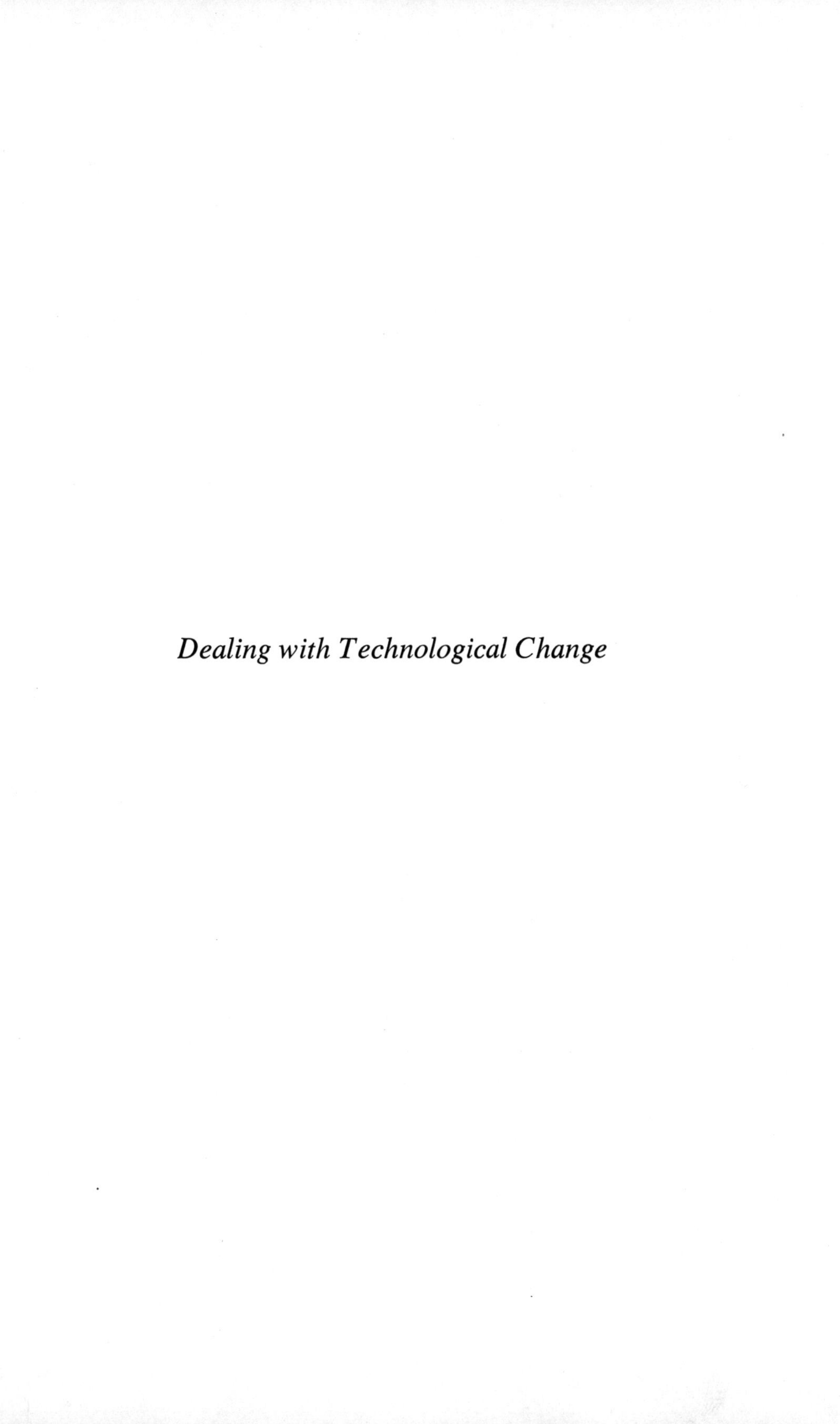

Dealing with Technological Change

DEALING WITH TECHNOLOGICAL CHANGE

Selected essays from *Innovation,* the magazine
about the art of managing advancing technology

AUERBACH publishers

princeton
philadelphia
new york
london

Acknowledgments are made to:

Gabor Kiss for the drawings of the planes in Chapter 2

Celesta Throndson for the map and corporate symbols in Chapter 3

Jeremiah Bean for the photograph in Chapter 6

Cal Bernstein from Black Starr for the photograph in Chapter 9

Published simultaneously in Canada by Book Center, Inc.

Library of Congress Catalog Card Number: 76-151234
International Standard Book Number: 0-87769-083-9

First Printing

Printed in the United States of America

Contents

Introduction

Whether you are trying to make technology work or whether you are trying to stop it from working—whether you are "protechnology" or "antitechnology"—this collection of essays is going to confirm some of your deepest fears and prejudices. Indeed, I believe this volume is capable of drawing you to either of two quite opposite conclusions:

- That technology is simply too all-powerful—and hence a danger to mankind, or

- That the introduction of new technology is so difficult, so costly, so fraught with risk, that it is a wonder new technology ever occurs at all.

Perhaps you think this is a warning. Perhaps you think the authors whom you are about to encounter are at war with one another over whether technology is good or bad, whether it is workable or unworkable. But not at all. In fact, there is remarkable consistency in this collection of viewpoints, given the fact that ten authors are at work here, telling us what they see in that land we call the technological world. No, this war is not bound into this little volume. The war is in your mind—you who hate and fear technology, and you who are its lovers. There is where the war is raging. And until you come to peace, until we all are wise enough to open our minds a crack, until there is enough enlightenment around to enable us all to see what enormous good technology can do—and what enormous bad—until then we shall continue to suffer the same old dreary irrelevant argument of vice versus virtue. As though technology were so simple as that. As though technology were spinach, and you could like it or hate it.

This book is a geography—a book of exploration. It is a book which takes you to technical places: the swampy land of autos, where technology chugs and sputters in the mud of long convention; the windy land of props and jets, where technology can blow you to oblivion; the powerful, surging land of electronics, where technological giants and clever pygmies sift gold from streams of abstract thought; the chasm of biomedicine, between the cliffs of medicine and engineering. These are the places—these and others—and if you are curious to see the sights and watch the animals eat dollar bills before your very eyes, by all means you should come along.

Each of these chapters appeared previously, in somewhat different form, in *Innovation*, the journal about the art of managing advancing technology. And each of the authors is associated with the journal as a member of its editorial staff or as a contributor to its mission of bringing understanding to the process of innovation itself. In either case, these authors are all grappling with the same phenomenon: man's ability—or his ineptness—at handling information. For that is all technology really is: knowing how to do something. That something might be the introduction of disk brakes in American automobiles, the sticky process which Donald Frey describes in the first chapter; or it might be the introduction of instruments, devices, and systems for health care, as recounted by James Dickson III later in the volume. But in all cases, there is enormous resistance to the new idea—enormous resistance to change—and oftentimes the resistance is so overwhelming that the idea is crushed by the environment to which it is exposed: as witness the aircraft fiascos in Warren Kraemer's Gallery of Magnificent Blunders, in the second chapter of this volume.

A century ago, Victor Hugo observed that nothing is so powerful as a new idea whose time has come. Certainly that observation held true for the time and the event of which those words were written—the time of the French Revolution. And today, in this time of technological change and upheaval, it might seem that the thrust of modern technology is so strong that, indeed, its time has come. But is that really true? Or is technology far from its time? Or even past it? I wonder if the course of our technology, the direction of its thrust, is taking us away from that point in time that could rightly be called an Age of Technology: to describe the 1970s as an Age of Technology is to describe the 1770s as an Age of Kings.

And that is a tragedy of our time. Never has man needed technology more. Yet, never has technology been so abused, so misapplied—never have we so wastefully developed tasks that do not need doing, while neglecting to pursue those that should be done. Some would blame technology itself for these things—for the fouling of our air and the plundering of our waters, for the neglect of hungry children and the blight of urban life. But the fault is not technology's. The fault is our own—in knowing how to do so many things and yet not even knowing how to order our priorities for human betterment.

David Allison

New Canaan, Conn.
January 1971

Part 1

Reactions to Change

1

The Colossus That Is Detroit

Donald N. Frey advises those "mechanics" who would tear down and rebuild the enormously complex machinery of the automobile industry (or any other major system). It is essentially the same advice he would give to the man at the corner garage: Tune it up —give it a major overhaul, even—but before you take it apart be sure you understand how it is put together.

Mr. Frey is President of General Cable Corporation.

One of the recurrent refrains of the late 1960s and early 1970s is the one that begins, "Any country that can organize its resources to go to the moon ought to be able to . . . (do such-and-such)." Usually, "such-and-such" involves a not-inconsequential change in one of the major system around which our society is built. Sometimes it is one of the large urban systems—say the city of New York. Other times it is the communications network, or the transportation network, or maybe that giant system known as the federal government. Sometimes it is the automobile industry.

What needs to be recognized is that while such-and-such can probably be accomplished, and may in fact be essential to the continued vitality of the society, it is by no means as easy to do as it appears. Making a significant change in the way the automobile industry goes about its business, say, may be as much as an order of magnitude more difficult than making the manned moon program work, and large changes in such complicated social systems as we have in the city of New York may be an order of magnitude more difficult than that.

Take just a very small event in the history of automotive innovations; take the introduction of disk brakes. Before they appeared on American cars people used to say, "Disk brakes have been on European cars for years. Why does it take Detroit so long to put disk brakes on a car? A simple little thing like that?"

What those people failed to realize is that Detroit long ago lost the luxury of having anything be simple. In the automobile industry even the simplest things are matters of great complexity. Once you decide to put disk brakes on the front wheels (just the front wheels, notice—you must start small) of the 1965 Thunderbird, let's say, you've made a decision that affects tens of thousands of cars, tens of thousands of people, and hundreds of manufacturers.

Before the brakes can be put on the car, you have to have the required manufacturing facilities in place, work out the assembly procedures, design tooling jigs, fill the parts pipeline, train mechanics to service the things, prepare, print, and distribute service manuals, prepare dealers and salesmen, set manufacturing and quality standards so you can make the brakes hour after hour, and hundreds of other big and little matters that someone has to attend to.

I was at Ford at the time this introduction of disk brakes was going on and I remember, for example, that we had lengthy discussions about the changes in shipping dunnage that would be necessary to ship disk brakes from the manufacturing plant to the assembly plant. We spent hundreds of engineering man-hours designing new ways of shipping the things, developing new packing geometries for putting them in a freight car, and setting new damage criteria to be sure they were not being bent, rusted, or whatever.

All of this takes time, perhaps five years between the time we knew we had the technology in hand to the time the brake was introduced on the market. And that completely leaves out all the earlier years spent in patient testing and development under an unbelievably wide variety of conditions of use (or, more properly, misuse and abuse, because that's what the product gets in the hands of the consumer). It also leaves out the years of research into high-temperature materials, heat transfer, friction coefficients, wear tests, and so forth. And in the end, it was the technology of the ventilated rotor disk from, of all places, the railroad car business, that made the whole thing possible, for the unventilated disk used on European cars is unsuitable for heavy American cars.

The point I am making is that when you manufacture a complex product in the volume that we manufacture automobiles, it is very difficult to make a big change in the production or support system, and even a little change— just the front brakes on one model—involves a multitude of not-so-simple changes on every level and division of the industry. This is not to say that Detroit always moves as rapidly as it can, or that it is as innovative as it could be. Far from it. It is simply to say that anyone who goes about to introduce change or innovation into such a large, in-place system as the automobile industry must understand the kind of system it is or his efforts will be in vain.

This whole thing has special meaning for me, however, because it seems to me that the automobile industry must be one of the classic examples of a huge, mature, established industry that cannot change easily, and yet depends upon change to stay alive. There are a number of other systems that have similar characteristics: the power generation and distribution system, the education system, the health-care system, the housing system. They are all ponderous, complex systems that have become an integral part of the economy. How change is accomplished in those systems, with the severe restrictions their size imposes, is, in many quarters, a wholly misunderstood process.

They all have certain qualities in common, these large systems. In the first place, they have a lot of money invested in them, which means you can't tear the system up and replace it with something new very fast. They also involve large numbers of people—both as participants and consumers—and that implies additional sluggishness because people's behavior patterns and predilections don't change very rapidly. And finally, whether it is apparent to the onlooker or not, there is usually a very sophisticated technology present, which means that you will have to have equally sophisticated technology if you want to change it.

Take the railroad business. Take their rolling stock. It's big heavy, clumsy, a lot of it looks as if it was built in the 1890s (some of it was) and as if it hasn't changed since. But do you realize what a pounding a freight car takes every year it is in service? The railroads beat the hell out of their equipment in service. A freight car is big and heavy for a reason: if it wasn't it wouldn't survive. I'm not arguing that technology can't develop a light, cheaper-to-operate, long-life freight car. I'm only saying that before you assume that modern technology could easily build a better freight car, be sure you understand why it was built that way in the first place.

All of my experience with these enormous, slow-changing systems has been with the automobile industry. It is, nevertheles, a strongly representative member of the family. It is, first of all, very large. If you take the whole thing, you must consider in addition to the automobile manufacturers and dealers, the road system, the corner gas station, all of the independent parts manufacturers and distributors, and hundreds of thousands of small companies whose business is dependent in some direct way upon the automobile. There is a ganglia of interconnected enterprises running through the whole American economy, and it goes far beyond the manufacture of those products that have Ford, Chrysler, GM, or American Motors nameplates on them.

Altogether, the automobile industry comprises about 10% of the American economy; it is said that one in every six people works in the automobile industry or one of its related activities. I added it up one time and discovered that the total investment in automobile transportation in the United States—rolling stock, plus roads, plus corner gas station, etc.—amounts to about 2 GNP's! That represents an almost inconceivable amount of inertia in what is basically a single approach to the transportation problem—individual, four-wheeled vehicles, individually powered and controlled. Whatever you do, it cannot be changed very quickly.

In addition, the automobile industry is unique in at least one respect: It produces a very complex product in very large numbers. The aircraft industry produces a much more complex product, but it makes 100 of them and sells them for $10 million apiece. The telephone company makes more telephones but, complex as it is, the telephone is much less complex than a modern automobile. In terms of volume and complexity—or let us say the product of volume times complexity—the automobile industry stands alone. There is no other human enterprise like it; in this respect it is bigger than anything else in the world.

We hear a great deal about the "economies of scale" that work to the benefit of any such high-volume operation. We hear, for example, that the manufacturing cost of an item drops at a predictable rate as the production volume increases, or that the cost of providing a given service falls in proportion to the number of people using it. We seldom hear about the "adversities of scale," the qualities that make a large organization harder to change, more difficult to keep viable and lively. Being big has its advantages, to be sure, but it also has disadvantages. The sword cuts both ways.

Let us suppose we are a group of legislators contemplating a law requiring that all automobiles be powered by gas turbines by 1975 so as to reduce the problem of automotive air pollution.

The first thing that happens is that someone does a little analysis and finds that a turbine is a very expensive thing to make, compared with the piston engine. But somebody else comes along who is a little more sophisticated and says, yes, but we will develop new materials and new production techniques and then the cost will come down because instead of making them by the hundreds we will make them by the millions. Economies of scale. At this point it begins to look almost practical, and that is frequently the point where analysis stops when it is being done by someone outside the automobile industry.

What gets left out in the analysis is that beyond the problems of designing and producing the things are the problems of marketing, servicing, fueling, and maintaining them. In the case of the gas turbine, probably the most important limiting factor is that there isn't the infrastructure in the United States to service a gas turbine at every corner garage. There are no trained people, no technology, no service equipment, no parts suppliers, no distribution for parts, no retail suppliers, no retail outlets for either the mechanic's labors or the parts themselves. You've got to turn the whole system upside down.

Even a small change can set into motion a remarkable interconnected chain of events. One of the changes I worked on in the last few years that I was at the Ford Motor Company was a switch from the traditional bias-ply tires to the new bias-belted tire. Bias-belted tires are now standard on all cars. One might assume that a change of that kind would be pretty easy to make: All we do is order a different kind of tire from our suppliers and put them on the car.

But there's that enormous system again. The story doesn't start with the bias-belted tire; It starts with the radial tire, which is the next step beyond the bias-belted tire in terms of departure from the traditional bias-ply tire. For many years we had been wanting to offer radial tires on our cars because they have a long wear life and provide good traction in the wet. We had been hesitant to do so, however, because they are expensive and can result in a harsh ride. But with the growth of consumer expectations and the pressure of the safety act we finally decided that we would offer them as an option of several car lines in 1966, as I remember. The reason was that the benefit outweighed the cost: We could recover our cost structure and the consumer got a better break. It was one of those cases where both the consumer and the manufacturer could make out.

We had to consider more than the cost of the tire when making this change, however. Because of the design of the tire, the suspension of the automobile had to be redeveloped, and that set off this chain reaction I have been talking about—engineering, tooling, service, manuals, parts supply, marketing—the whole thing. So it was not a minor change for us. As it turned out, it was not a minor change for the tire companies either.

When we developed the suspension to accept radial tires (and incidentally it still had to be able to accept conventional tires as well) and offered radial tires as an option, we had no real idea of the size of the market. We didn't know whether we would sell 1% radial tires or 10% radial tires. We simply stocked the assembly plants with every radial tire we could get from every manufacturer we could get, despite the fact that they were available in only small quantities at that time.

This initial step showed we were serious about improving tire technology for our customers, and the customer responded by buying more radial tires than we anticipated. But that put the tire companies in a bind because they didn't have the tooling to manufacture large numbers of radial tires and you can't build radial tires on ordinary tire-making equipment. Therefore the tire companies turned around and developed the intermediate tire—the bias-belted tire—which can be built on the existing tire-making machines and can be produced with much less capital investment. It is my belief that they will build up their investment in radial tire-making equipment over the years and eventually the radial tire will replace the bias-belted tire as a volume premium tire.

This is typical of the history of innovation in large systems: A small change here triggers another small change there, which sets off several other changes elsewhere, and pretty soon you've turned the whole system around. It can't be done overnight, and it can't be done without understanding how the system is set up. But it must be done, and despite increasing difficulty, it will be done.

One of the important characteristics of large systems is that innovation on the technical side demands innovation in other facets as well. The engineer's Hell is paved with brilliant technical innovations that failed because the company never solved the marketing, distribution, and service problems. They didn't understand the system they were working with. Somebody was hollering down a rain barrel and all they got back was an echo.

People generally think of innovation in terms of technology, science, research, and engineering. When they talk about introducing innovation they are generally talking about technical innovation. It is rare that anybody talks about technical innovation in terms of figuring out a new way to market the stuff. Perhaps it doesn't market under the old rules. Perhaps you can't run it through the old distribution system. You can't put gas turbines in cars next year or the year after that, not because we haven't solved the technical problems (although those are difficult enough) but because we haven't got the system to take care of them. You think you've got service problems with your automobile now; put a gas turbine in every one of them and see what happens for the next decade.

There is just as much risk in introducing novel new ideas in marketing or service areas without first understanding the system you are dealing with. Let's take service innovation. Suppose we decide to set up service centers that are not tied to the retailing of new and used cars. It's a suggestion that has been tried in a few places. Will it work? Is it economically viable? First you must ask if the service business is there only because it is a necessary adjunct to selling cars, and you have to have the two hooked together to support the service business. It may turn out that the two can be separated, but first you must ask that key question.

Service how? By unit replacement with a new one, or by on-the-spot repair of the old one. It may seem more practical to merely require the mechanic to replace a defective black box with a new one—a new carburetor for a faulty one, for example—but it may not be economical. You come up to the question of what do you do with the faulty carburetor. Do you send it back to the manufacturer and, if so, what are the economics of *that* system? It has turned out in a couple of cases that despite the beautiful theory of unit replacement the economics simply wouldn't work.

Or let's consider franchising. Suppose you decide to separate the automobile franchises from the heavy truck franchise; there's no sense in having the big over-the-road trucks sold at every Ford dealer in the United States. So you decide to specialize the franchise. That's an innovative step taken by a former boss of mine. But it is an enormous one because you have to get all the dealers to agree to give up their heavy-truck franchise. After all, they have legal rights under these franchise contracts and the automobile company can't simple change the agreement unilaterally.

Or let's look at warranties. I noticed in the paper recently that Ford has reduced the warranty period on their automobiles. That's a big step. (Some may believe it is a step backward.) One of the factors in that decision was that the FTC (and the public) has been complaining recently about the inequitable administration of warranties, the deception of the consumer, and so forth. It is interesting to reflect upon the fact that one of the solutions to that problem is to simply eliminate the warranty. That's one solution. I'm not sure it is the one the FTC had in mind, but it is one solution.

I'll say it again: Before you introduce an innovation you must understand the system—understand among other things that it is frequently big and complex and slow to move. Then you are in a position to take a small but significant step. Until then you are a neophyte, you are innocent, you are naive, and you will get bloodied up and thrown out.

Change is essential to the system, and yet the bigger the system, the less important change seems to be. The big system seems to have a stability of its own. Moreover, the effect of the small innovative step you are about to take seems as if it will get lost in the big system. The short-term risk looms large by comparison with the amount of difference it seems to make.

The beautiful hidden problem here is that if you don't innovate, if you don't take what you think to be the short-term risk today, you are taking the biggest risk of all, which is the risk of becoming obsolete in the long term. It's an easy mistake to make because the long-term risk of obsolescence is always tomorrow. It isn't in front of you every day and it isn't as well defined as the short-term risk. It's the problem of the risks I know today vs. the risks I don't know . . . well, I may even be retired by then.

It's an article of faith in the automobile industry that you have to change or you will die. Change includes not only technical changes like changing the brakes from drum to disk, but styling changes, and changes in structure of the organization or the way of doing business. Change is inherent in the automobile business.

The pressure for change comes from all quarters, but by far the most important one is competition. About two thirds of the new cars purchased are the same make as the last one the customer owned. Now, that's a high degree of brand loyalty, but it still leaves a very large number of customers who presumably have no preconceptions about the brand of car they want and are possible targets for any of the manufacturers. Every manufacturer wants as many of those uncommitted customers as he can get, and he depends upon innovation to attract them. He attempts to offer a more attractive package with more desirable features with more appealing marketing than his competitor offers.

But his first responsibility is to hang on to the two thirds who purchased his company's car the last time, and there it is not so much a question of what he offers this time as it is a question of what he provided last time. Basically, that's a question of reliability: It didn't leak, squeak, rattle, or fail to start every morning. And the gas and oil consumption was, in some mysterious way, acceptable. Such reliability is also a matter of innovation and, again, it doesn't depend upon a single feature, such as a transistorized ignition system, or even upon a single kind of innovation, such as technical innovation, but upon the whole range and scale of activities that characterizes the automobile industry.

It isn't enough that the car is designed to be reliable, it must be put together properly, it must be prepared for delivery properly and it eventually must be serviced properly and expeditiously. It does no good for a manufacturer to design a trouble-free automobile if his service representatives don't have the proper tools to maintain it, or the proper parts to fix it when it does misbehave. A man's view of reliability depends on a whole spectrum of things: Did he have to call in for an appointment three weeks from now and get up at 7:00 in the morning to get his place in line? Was it fixed right the first time? Was the dealer courteous? Was the car out of service a few hours, or did it take weeks? Holding on to his two thirds requires of the manufacturer as much innovation—probably more—as it does to attract that uncommitted one third. But it's a different kind of innovation.

It is typical of large systems, and the automobile industry is as representative of this as any, that the targets do not stand still and that what is good enough today is completely unacceptable tomorrow. With the passage of time, innovation grows more difficult, more expensive to accomplish, more time consuming to introduce, more risky to invest in. This is partly because the system grows inevitably bigger, but it is also because the pressures on the system begin to multiply and begin to push it in unaccustomed directions. What began as a large, rather complicated system responding to a rather simple set of forces becomes an enormous, exceedingly complex system responding to a variety of simple and compound forces.

Consider cost. One of the first things a junior engineer in the automobile industry learns is that his great idea for improving the product is fine and right but, oh, by the way, it shouldn't cost any more than the last way we did it. In fact, in most cases, the last way we did it is already too expensive and we are already looking for some new way to do it at less cost. Someone who worked it out not too long ago discovered that if you made cars today entirely the way they were made fifteen years ago, but with today's labor rates, it would come out to something like twice the present cost. In that period of time, we have probably reduced the direct labor cost per car to about half of what it was. That wasn't done at no cost. There was, first of all, the cost of an enormous engineering effort—simplifying fasteners, reducing a three-piece assembly to one, replacing a machined part with a formed part, thousands of big and little things that add up to a product that costs less to manufacture. It was also at the cost of a very large investment in capital equipment—automatic machinery that allows parts to be made without the direct labor of people.

One of my favorite examples of this is the speed-control device that Ford offers as an option on certain of their automobiles. It is sold in some volume, not much, perhaps tens of thousands. What appeals to me about it is that it is a beautiful little servo, with all the hydraulic, mechanical, and electrical technology of the servos that are used by the dozens in airplanes. If you described it to a group of engineers, just in terms of its performance characteristics, you would talk about it as a closed-loop, closed-center-valve servo system, with such-and-such a response and such-and-such a transfer characteristic. It could easily sound as if it cost hundreds and perhaps thousands of dollars. Surprisingly enough it has a direct manufacturing cost of, as I remember, about $18. The first hand-made one probably cost hundreds of thousands. And then it got down to $1000. And then the process engineers went to work on it and spent thousands and thousands of man-hours. "Do I have to do it this way? Can't I do it that way?" "Instead of a fancy investment-cast, alloy-steel part, I think I'll make it on a punch press." And pretty soon they are making it for $18.

The cost restraint works both ways: on the one hand it discourages the good, but expensive idea and on the other hand, it stimulates innovation aimed at reducing cost. On balance, it is a much greater stimulant than it is an inhibitor.

In recent years there have been new pressures on Detroit, and like the restrictions of cost, these pressures have both an inhibiting and a beneficial effect. These pressures come largely from government, but also concerned citizens' groups troubled about such issues as highway safety, or the polluting effects of the internal combustion engine.

On the beneficial side, the new pressures have the effect of establishing a constructive set of tensions between the government and the automobile industry. For the most part, the industry has to be pushed into safety and anti-pollution reform because their first concern is to sell a product and make a profit, and these modifications generally raise the price without necessarily making the product any more salable. So there is a provocateur role there, and I don't think there is anything wrong with that.

Consider the safety issue. There is continual disagreement between the automobile manufacturers in Detroit and the administrators of the safety program in Washington. This is to be expected: There are two different points of view and two different sets of equities. From the point of view of the administrator in Washington, the question is how to save lives by changing the design of the car, and from the point of view of the auto manufacturer, the question is what the design will cost and whether he can sell it to his customer. Those are the elements of a constructive set of tensions. Out of these intense encounters comes practical progress.

On the negative side, the pressures from the government increase the business risk connected with the introduction of new safety devices. When you change the braking system from drum to disk, for example, there are bound to be difficulties in the beginning, and if the government requires that cars be publicly recalled under conditions that may unjustifiably discredit the manufacturer or the design, the net effect will be to inhibit manufacturers from trying something new.

Or let's take the air bags that have been proposed as an alternative to seat belts for restraining passengers in a collision. Ford has been running some preliminary tests on air bags over the last few years, and now the safety administrators have begun to talk about requiring air bags on all cars by 1972. That could be a very serious problem for the automobile industry because the equipment is not yet ready either from the point of view of fail-safe reliability, or from the cost/price/value point of view, as measured in the minds of the customers who will have to pay for it. If Detroit were unable to meet either requirement by the deadline, it could have very serious consequences. And so one of the new risks of introducing a new idea is that somebody will turn around and make it law before you are ready for it. I see that risk getting larger every day.

But there is no question that the principal effect of new pressures is to produce better automobiles, not by the imposition of a set of design rules, but by focusing public attention on a neglected area of design—safety, for example. In that way, the people who should be thinking about better ways to make automobiles—the professional automobile engineers—will begin to think about safety. Safety then becomes a competitive factor in the marketing of automobiles. That's what the safety administrators really have going for them. The result of that set of forces—the competitive forces—will probably be the biggest single factor in the improvement in automobile safety. Given that incentive of self-interest, watch what Detroit does with it!

The constant risk with a large, slow-moving system is that it will not be able to respond quickly enough to large shifts in technology or consumer demands. That is when the system becomes obsolete, and you become the last manufacturer of ice boxes instead of the first manufacturer of electric refrigerators. The shift may come from within the industry, or it may come from outside; it may appear quite slowly and subtly, or it may come on quite suddenly and with obvious implications.

Today, for the automobile industry, there are warnings that the internal-combustion engine is about to be displaced by some other kind of power source, or that the entire concept of individual private automobiles is going to be scrapped in favor of some kind of mass transit system. Such warnings cannot be ignored. Such changes could very well occur, and it is essential that the people in the automobile industry understand the implications and probabilities of such a change—and they largely do, I think.

The company's early warning system for detecting the approach of revolutionary technical change, as well as smaller changes, is the research department. It is the research department that should give management the foresight needed to appraise future developments and future changes in the market. And it is research that keeps the company in touch with a wide range of activities, both within the industry and outside of it. It assures management that they have cast their net widely enough to catch the broadest possible variety of ideas.

It may seem strange to describe research as information gathering rather than as exploration, but it seems to me that this is as important as any contribution it can make to the corporate scheme of things. A new idea, a new way of doing things, is as likely to come from someone else's research organization as it is from your own. It is as true of other industries as it is of the automobile industry. The research done in your own labs is merely a contribution to a vast technical base. What the company takes out of this base may have nothing to do with the basic research it put in. There is no one-to-one correspondence. The research investment is, as much as anything, an **admission** ticket to a club of the world's scientists. You don't depend on it for your commercial ends to any degree. You depend instead upon the world's technology, to which you have access by being a member of the club.

Basic research also provides a trained group of observers and interpreters of what is happening at the forefront of science. Such people cannot be onlookers of science, they must be participants. Researchers are the best qualified to understand the importance of anything that may happen in their field of technology and to appreciate the significance of it as it relates to your business. And again, it may have nothing to do with what's going on in your laboratories at any point in time.

Another factor that is going to help keep the automobile companies from putting themselves out of business is the new emphasis on what has been widely called systems engineering. It doesn't matter much what you call it, but what it amounts to is that the automobile companies are beginning to concern themselves with a much wider range of activity than the narrow concerns of building today's kind of transportation (and at a profit). At about the time I left Ford, the whole organization was being set up according to the systems concept. It was not revolutionary change; the organization had always been structured around the systems, subsystems and sub-subsystems of the automobile. I think the change was more a reflection of the fact that the automobile and its various subsystems are becoming more complex and more technically interrelated, often with things outside the automobile.

Forty years ago internal combustion automobile engines were producing an amount of pollutant that was unnoticeable in the atmosphere. Today, sheer numbers have made pollution an important factor in Los Angeles. Suddenly the automobile manufacturer has to relate to the agricultural economy of the area—he is messing up the truck farmers, not to mention the citizens who get tears in their eyes. What it means is that he must design automobiles for something more than safe transportation at low cost; he's got to worry about his effect on the ecology. And that, in modern parlance, is a systems problem.

Or consider another example. With the advent of skid control on automobiles (and I understand it is about to be standard equipment on the Mark III Continental) you've got a little black box full of electronic logic under the glove compartment to control the skid condition on the rear wheels. With that small step, I could take another step and connect the logic box to an external control that will intercede in the logic and cause the car to stop in response to some kind of roadside signal. Suddenly I have a problem that extends beyond the traditional boundaries of the automobile manufacturer. Suddenly I have a systems problem. Who's going to provide the roadside signals, and for what purpose? What about compatibility with cars not equipped with the stopping equipment? What about the cost distribution between what you put on the car and what you put on the signal and who is going to pay for what? There are maintenance problems, political problems, legal responsibilities.

Detroit long ago passed the age of innocence; it is a big system and no amount of simplifying or restructuring will change that. The concepts of systems engineering—regardless of what you call them—are an essential element of the modern picture.

For some, this is cause for distress. The big organization seems so much less manageable, so much more likely to get out of control, so much less responsive to the usual management techniques. Contributing to the large organization may seem much less rewarding than working in some organization where dramatic changes can be brought about almost overnight.

Dramatic changes, however, are not the stuff of which technology is generally made. Nothing we did at Ford was revolutionary: nothing as spectacular as the development of the transistor, or atomic energy, or the computer. But those are the exceptions. The bulk of technology is slow, evolutionary plodding.

I like to think that it is just as intellectually challenging—and rewarding —to contribute to this slow metamorphosis of a massive system as it is to play a role in a revolutionary development. To take one of these big, big systems and change it—that is the workaday world for the technologist, and a far more common thing for most of us.

October 1969

2

The High-Technology Trap: Too Much, Too Soon

The U.S. aircraft industry is on the brink of turmoil. One giant firm has toppled, a second may be in grave danger. Poor management? The answer is more complex—and its significance reaches far beyond the aircraft industry. Every technology-based industry faces the same danger, for each has the same problem of adjusting to radical technical changes.

Warren E. Kraemer has been living with one industry's efforts to make such a transition.

Mr. Kraemer is a top officer with Scandinavian Airlines System.

My industry, the aircraft industry, is steering a course toward trouble.

It may surprise you that I should play Cassandra just at a time when big orders for new jumbo jets and airbuses seem to have pulled the industry out of a threatening nose dive. But some companies did come close to a crash. What worries me is that although they know they just pulled up in time, it is still not at all clear to them how they got into such a predicament in the first place. Now there are economic downdrafts ahead for jumbo jets, as well as technological turbulence for the SST, but few companies know it—yet.

Paradoxical as it seems, major improvements in the technology of aircraft production might well be blamed for the companies' troubles.

Dazzled by its new techniques, I suspect, the industry lost its sense of balance. And as I shall show later, the very nature of the industry sowed the seeds of subsequent troubles into each apparent success.

Still, it did look like good flying weather for the industry a few years ago when it became clear that extremely complex aircraft could be produced in astronomical quantities. At that point the manufacturers decided on a design of the complex sort they could obviously handle. The aircraft would be jumbo size and would have a remarkable performance spectrum spread: minimum approach speeds and maximum cruise speeds, minimum balanced field requirements and maximum payload ranges. They peddled this kind of design hard enough to talk their customers into signing up for a production run.

That was their mistake. The manufacturers tried to exploit what they could build—a radically new generation of aircraft. Instead, they should have thought out passenger and cargo requirements, in the air and on the ground, and adjusted to the needs of the operating environment. They should have considered, too, the sensitivity of timing when inserting a major innovation into an existing spectrum of technology then barely beginning to be amortized by its users.

Better ways to manufacture don't always lead to market success. Just a couple of years ago the electronics industry, spurred by new ways to fabricate and integrate circuitry, went through the same kind of trouble.

It is this *kind* of trouble I want to describe. For I think some important lessons were learned in the electronics industry case—and along with good luck and good management of technology, the lessons may enable the deceptively booming aircraft industry to save itself before it slips into another spin.

These lessons may also save some other industries trying to cope with the subtle effects of changed production technology and the often unwise decisions one makes in a blind effort to constantly feed an expensive elephant in the barn.

By and large, the crucial decisions of how and what to feed the elephant are made by technology-oriented men. These are men responsible for vast laboratories and plants filled with people who, after much tortured exposure, have mastered the ultimate in their particular sphere of skills and learning. These range from specialists in riveting, or threading electronic harnesses, or connecting complex hydraulic plumbing up through layer upon layer of supervisory shop personnel to the highly-paid project engineers and managers possessing advanced degrees in electronics, metallurgy, or chemistry.

Still, these men, by and large, do not *initiate* anything; they are cogs in an enormously expensive design and production machine with a capability both defined and limited by the sum total of their technological expertise. The whole thing is an elephant, a monster. Still the very existence of this monster in the barn is almost impossible to ignore at the planning level: Can the planners tell management to scrap a whole technical approach and start over again while the payroll eats the company out of house and home?

On the one hand, if the planners fail to provide work *and* challenge, the team may drift apart, attracted by the blandishments of other companies fattening their technical teams for competitive technical bidding or actual contracts. On the other hand, the team must not be permitted to force adherence to its own line of thinking, extrapolating past capabilities into presumed future successes.

Characteristically, the judgments of a mere handful of decision makers can be polarized toward feeding the elephant in the barn. Only the diet might be wrong.

First let me describe some characteristics of the aircraft industry which tend to breed this kind of trouble. Unlike manufacturers of the automobile or similar consumer goods, the final arbiters of success or failure for this industry are not among the buying public, but consist of a relatively few key men who will purchase its products to perform certain services, in defense or in civilian transport.

Unquestionably the industry has been overwhelmed by its own track record. Riding on conviction and presumed expertise, airplane manufacturers and engine manufacturers have demonstrated that they can join forces to sell a contract to the military, or to bring a concept to the air carriers.

For example, at the time the contract was let by the military for the giant C-5A cargo and troop-carrying airplane, the forerunner of the jumbo jets, the Air Force's need for it was undeniable. There would be problems in designing and building anything so unprecedentedly large, but all during the period when the manufacturers were chewing on the problem, it was no secret to the air carriers that the development of a next-generation airplane would soon be underway. Even so, the carriers evidenced very little interest in what this new airplane might represent to them in a commercial sense.

It was not until Boeing lost the competition on the C-5A contract to Lockheed that the assault on the civil aviation market began—and here it begins to become apparent that the aircraft industry does not adjust its capabilities to customer need, but sets them according to estimates of its own potential for growth. In Boeing's case, the company had just lost a competition which involved a huge expenditure just for it to enter. In terms of working capital, cash flow, and disposable asset posture, it had sunk about ten million dollars of its own money.

From an administrative standpoint, Boeing is a beautifully organized company. It didn't take too long to decide that it could convert this loss to a gain if it could apply the technological input it had made in developing its proposal for the military C-5A to building its position in the commercial transport field. The ink was hardly dry on its competitor's contract for the C-5A before Boeing had made some pen strokes across the very complex variable inflation landing gear (required for soft, forward military air fields), changed some of the supersensitive onboard electronic equipment like peripheral search radar, and altered the flow-through loading (in through the nose and out through the tail) to more conventional schemes of access. The whole 747 was then scaled down in weight and size to a 680,000-lb. level from 814,000-lb. maximum all-up weight of the C-5A.

Then came a magnificently mounted sales appeal. It alluded to the golden harvest the airlines had been reaping from Boeing's spectacular success with the 707 family, born out of aerial tankers designed to refuel B-47's, and it offered a new fallout to the air carriers: the 747 jumbo jet.

From a technical standpoint, it could appear that the 747 was part of the same aircraft family—simply expanded in dimension, lift, and cubic capacity. The number of men in the cockpit would be the same as in the 707, and it mattered little if the captain got a raise from $38,000 to $50,000 per year; after all, he would be flying 500 passengers at a time rather than 186.

The carriers were swept off their feet by the supposed economic gains they would sustain in reduced operational costs per seat-mile. They succumbed to Boeing's proposition long before they had thoroughly analyzed the aircraft, its systems, its application to routes or markets or ground facilities.

If there is a question of responsibility in the obligation to understand the economics of new technology, then perhaps both seller and buyer were too hasty. The aircraft manufacturer especially is bound to a huge research and production operation in such a way that it controls him. His elephant in the barn is a huge research and development operation and an enormous production capability, both manned by highly sophisticated technical personnel.

To keep the corporate structure and capability from dissipating, to maintain it as a productive organism, requires contracts and orders. It is particularly true in the aircraft industry that if you want to go out to get government business, you'd better have the engineers and the plant for a working team to back up the bid.

At the time we are talking about, the C-5A was the only major military work available, and Boeing had lost it. The supersonic transport was years away from production. Boeing was under the enormous pressure of a staggering weekly payroll. And company officials faced the horrifying specter of the Douglas Aircraft Co., which had literally gone out of business because business was too good (a paradox which I shall explain later).

Small wonder that Boeing tried to create a market by hurriedly training its sights on the air carriers, even though the company was delivering present-generation jet fleets to these same customers.

To the air carriers, what Boeing was saying sounded pleasantly familiar: "If you operate a bigger, faster airplane, *ipso facto,* you make more money." Stretched versions of the 727, the 707, and the DC-8 were already demonstrating that seemingly universal truism.

But it was on precisely that point—where technology and economics interface in a moment of glorious truth—that the carriers and the airplane manufacturer fooled themselves in two critical ways.

First, there was the illusion that the 747 jumbo jet was a current-generation aircraft. Certainly it looked like most other subsonic jets, with its swept-wing design. But its differences far outweighed the similarities. The most striking difference was pinpointed by the engines.

To power such a large new aircraft required a high-thrust engine. What Boeing expected to get was an engine capable of delivering a significant gain over previous designs on its thrust-to-weight ratio and specific fuel consumption throughout its entire altitude spectrum of performance. Based on this expectation, the airframe manufacturer had calculated that his aircraft could operate at significantly higher altitude level, thus gaining speed and reducing fuel consumption rates by virtue of the thinner air.

As it turned out, the designers got the expected takeoff power and climb rate, but then the figures began to decay. By the time the aircraft reached 35,000 ft., it would be at its most economic level of operation for its planned speed. However that altitude is the approximate operating level for the current-generation jets, which meant the 747 would become a prisoner of the air traffic control system. Airways control simply could not let aircraft, dispatched on a nose-to-tail basis within an assigned block of airspace, override the other aircraft ahead.

In other words, the 747 would be unable to operate economically at the higher altitudes which would give it a fast lane of its own. Consequently much of the proposed economic rationale for the 747 began to backfire. For example, the sweepback of the wing had been designed to be significantly more acute than on current-generation jets because the standard operational cruise environment assigned would not be around Mach 0.82 but Mach 0.88. The wing might well have to be redesigned because the engine planned for it is still years away from optimum performance.

All of these problems can be expressed in mere tenths of one cent per offered seat-mile in direct operating cost. Now this may seem like a mighty struggle merely to optimize minute levels of gain. But the airline industry

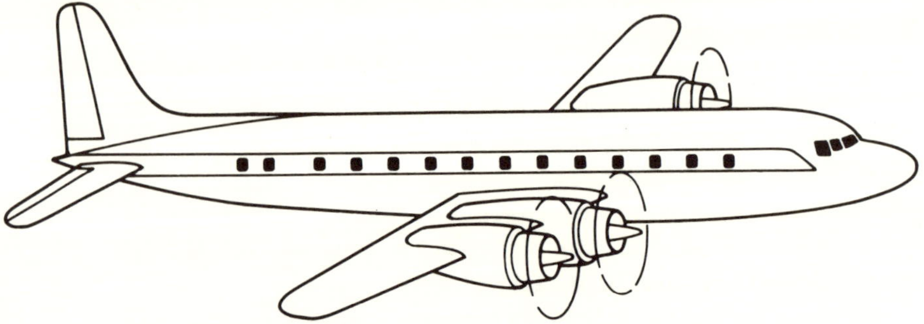

The **Lockheed Electra** entered the spectrum of aircraft technology during the transition from piston engines to pure jets. The designers tried to exploit the low cost capability of turboprop engines for short hauls, but competitive turbojet aircraft turned out to be cheaper to operate for short as well as long hauls. Worse still, the angle at which the engines were mounted on the initial Electra fleets caused severe secondary vibrations, overloading the wing structure. Tragically, wings fell off in a number of accidents before Lockheed recalled all the aircraft for expensive refitting. This put the company out of the commercial business for years. Ironically the same technical problem had plagued a much earlier transport—the first Electra.

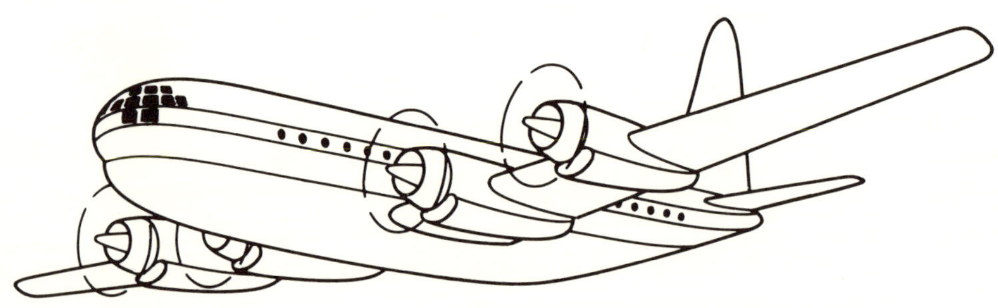

The **Boeing Stratocruiser** was loved by passengers for the contents of its spacious double-deck cabin with a cocktail lounge in its belly. Unfortunately for the airlines the aircraft couldn't operate at a profit because it quite regularly developed engine trouble; the turbo-compound engines couldn't deliver the necessary power without strain. United Airlines kept filling up the seats on its Hawaiian route, but never made a profit on that run until the day it stopped flying the Stratocruiser.

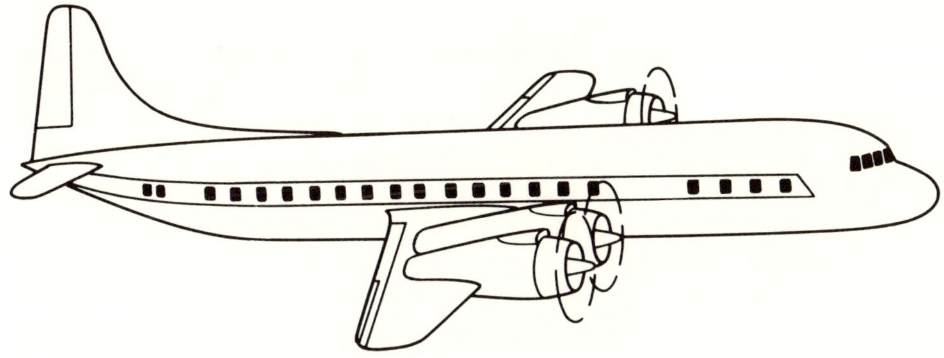

The **Douglas DC-7** was one of the last of the great piston-engined transports. Perhaps it never should have been built, for the airframe weight required a lift capacity and range from engines that were almost beyond the state of the art in the mid 1950's. Nighttime passengers sitting over the wings had to be reassured by cockpit announcements that the sudden brief silence was a shift to high-blower engine operation and that the glowing engines were not really aflame. Under continual strain the engines frequently did fail. When they did work they rarely were able to continually deliver the speed or altitude necessary to fly this aircraft at a profit.

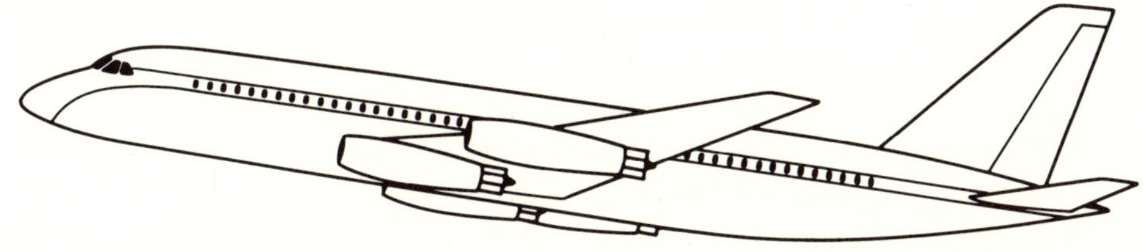

The **Convair 880** was a monumental gamble to attract a market with an aircraft reaching the upper limits of subsonic speed. It was an airborne hot rod, structurally a beautiful plane, but turned out to be an economic blunder. To gain the speed seemed to require a thinner fuselage, leaving one less seat per row over the entire length of the cabin. This clipped 18% off its earning capacity, too much of a penalty to pay for just a bit more speed. General Dynamics lost so much money on this plane that it almost went out of the commercial transport business.

operates billions of revenue passenger-miles per year and has to deal in a thin shading of gains.

For example, Eastern Airlines closed out its 1968 operation with about $744 million of revenue. There was a $7 million operating profit but a $12 million net loss after debt service. You can imagine how great the pressures to go after tiny increments are. A rough calculation shows that saving twenty seconds per telephone call would, in terms of Eastern's 85,000 calls per day, equal 170,000 man-hours (or girl-hours) per year in saving.

I cite this example not as an aside, but because it is typical of an airline operating environment which the aircraft industry, as supplier, overlooks all too often. And it is in this context that the jumbo jet manufacturers are now trapped. Not only are the airports and terminal facilities not yet ready to accommodate these huge aircraft and their enormous passenger loads, but some of the mechanical ground services being designed for them seem destined, inadvertently, to immobilize or incapacitate a multi-million dollar aircraft.

When the electronic baggage sorter and conveyor sticks—the aircraft is just as stuck. And even now an aviation insurance underwriter turns green at the mention of forklift trucks which are often used for ground service functions. A $60 a week ramp agent accidentally attacking a $7 million airplane with a $5,000 forklift truck can wipe out a great deal of revenue.

With the increased capacity of each jumbo jet, and the smaller number of them used, the immobilization of one aircraft can be even more costly. It seems likely that a major trouble for operators of the 747 may come from an over-the-wing loading and unloading device—a bigger, heavier, more complex version of the extendable jetways now used at some airports. As happens with even simple things like self-contained stairways on today's jets, sooner or later such a massive hunk of iron will become stuck in position for one technical reason or another and the aircraft will be delayed on the ground instead of earning revenue.

However, it will take no accident or mechanical failure to create an adverse operating environment for the giant new products of the aircraft industry. The airport environment, the air traffic environment, and the market environment have been plainly visible for years—and the first two have been relatively static.

The introduction of anything new into such environments is exquisitely sensitive to timing. I have already spoken of a saturated air traffic system and the inability of the forthcoming jumbo jets to operate economically at the higher altitudes where they would have fast lanes to themselves. By the same token, the jumbo jets will be involved in and will create traffic problems on the ground by virtue of their huge size and handling complexities. There aren't enough gate positions at most airports to handle present-generation jets.

The operating environment—in the air and on the ground—is clearly a factor in the total economics of the introduction of new aircraft, yet the aircraft industry has concentrated its marketing efforts around the cost of its products per offered seat-mile. The airlines are at last beginning to discover that they really ought to look at *all* the costs that are involved with an aircraft. Many a corporate head of an airline today who has placed an order for a 747 fleet would be delighted to hand it over for a more viable substitute. The headaches he is having stem from an economic aspect of the operating environment which I shall now describe.

The primary headache has come from the delusion that there are cost benefits to be derived from ever-increasing vehicle capacity. When the technical decision makers of the aircraft industry decided it was indeed possible to build a Queen Mary of the sky, there was hardly a pause to consider whether that large an aircraft might have to be used more like a Hudson River Dayliner.

True, only a few more crew members are needed to carry a vastly increased load of passengers in the cabin of the 747, thus lowering cost per seat-mile; but this hardly matters unless the market is there to fill the aircraft with great regularity. Though the airlines have touted the idea of more room per passenger, and lots of lounges and theaters, making the jumbo jet a great ocean liner of the sky, no treasurer of an airline, no comptroller, no banker will permit that airline to ignore the potential cubic capacity of the aircraft. When it's high season time and the passengers are being turned away, the potential extra one hundred seats surely will be installed. Only a few airlines, like National and Pan Am, are fortunate enough to be able to lease aircraft back and forth to each other because their high seasons come at different times of the year.

But what of the rest of the year? Then the capacity of the aircraft is too large for a market spread broadly but thin. That's when it can prove better to spread the services of smaller aircraft over the market—aircraft able to earn their keep with smaller loads and to provide more frequent and flexible service.

I am not against the concept of the jumbo jet in itself. What I am really talking about is the deployment of a new technology—jumbo jet or any other —without careful thought about the *all-up* costs to the user.

In the case of a new aircraft, these all-up costs are in contrast to the so-called direct operating cost of flying an aircraft between two points. Typically all-up costs add 100% to the direct cost. In the case of the jumbo jet these will be proportionately higher—both out-of-pocket for jumbo-size stairways, starters, tow trucks, access ladders, etc. plus new fixed terminal units for passengers and cargo.

The first jumbo jet, I insist, is an example of acting on the basis of what is technologically possible—feeding the elephant in the barn—instead of making a judgment in the light of operational environments and market timing.

Here I must emphasize the distinction between the 747 concept and the almost as large airbuses subsequently offered by Lockheed and Douglas. I am not happy with the word airbus, yet it seems to have become a popular way of designating large aircraft with high-density seating. Like the 747, Lockheed's L-1011 and Douglas' DC-10, though both three hundred seaters, are next-generation aircraft, but it's clear that a keen analysis has been made of how technological capability was to be utilized in them for more effective products in the market.

Perhaps these companies profited by Boeing's experience. The Douglas and Lockheed designers appreciated that the 747 was too large an aircraft for today's transportation market. As I said above, it makes more sense today to offer less unit capacity and more frequent service than to gamble on assembling a concentrated load at highly arbitrary departure times. So the new airbuses are designed to handle nearly one hundred passengers more than the current, stretched out four-engine jet aircraft, but still significantly fewer than the 747.

Furthermore because so much of the operational economics of an aircraft is involved in its engines, the tri-jet airbuses can save almost $5 million per year simply by having one less engine. Not only is there less fuel consumption, but one less engine teardown, inspection, and build-up which each time costs 25% of the acquisition price of an engine.

The costs of reliability and maintenance have serious consequences upon another factor which seems, at first thought, to have nothing to do with the operating environment of an aircraft—the market for used airplanes. The resalability of aircraft in secondary markets is as significant to the aircraft industry as is the resale value of automobiles to Detroit. Just as it is hard to sell certain brand-new automobiles today because they seem likely to have little resale value, so can certain kinds of aircraft prove hard to sell. The final generation of large propeller-driven aircraft could hardly be given away, even as freighters, because the operating costs were prohibitive when compared with costs of running new jets.

But the real lesson is still to be learned from this—that the costs of operating even a new-generation aircraft like the 747 can destroy its resale value in secondary markets. At first it was thought that the huge 747 might later continue to be used as a cargo craft. But subsequent economic analysis discloses that its payload, range, and the handling environment it needs—the cargo hatches, loading system, curvatures and structural support—all militate against its use as a freighter without major modifications.

Slowly but surely, the penalties for not looking at *total* operating expenses are being learned by the aircraft industry's best customers, the airlines.

It would seem that the manufacturers who survive will do so by introducing only the most effective, efficient, and economic aircraft into the existing spectrum of technology and by doing it at an appropriate time. Yet if that were the case, the great Douglas Aircraft Co. would not have gone out of business.

Today Douglas is a division of McDonnell—submerged into another company because its business was *too* good. Douglas went out of business because, at a point in history, it correctly identified the need for a modern, high-performance twin-jet transport.

With accuracy and with vision, Douglas designers responded to the requirements for a high-density, medium range airliner with low plane-mile costs (not just low seat-mile costs). By carefully matching power capability to gross weights, they substantially reduced operating costs—using two engines instead of three, just as the tri-jet airbuses will use three engines instead of four.

The orders rolled in for this superb design—you'll recognize it as the DC-9—and Douglas geared up for production, oblivious to certain danger signs.

Here I must explain that no aircraft manufacturer does all of its work in-house. Few make as much as 35% of the total weight of the airframe. Douglas was no exception to this rule.

Douglas' problem was that right then, the aircraft industry was booming. All the well-financed subcontractors had been siphoned off by the demands of other manufacturers. When Douglas found subcontractors, they turned out to be companies that were, for the most part, underfinanced and had to pay cash on the barrelhead for everything.

The mighty Douglas Co. had been in business for forty-seven years and was then grossing about $2 billion. It needed the subcontractors if it was going to fill all of those orders, and so there seemed little wrong with patting the contractors on the back and telling their local suppliers to send the bills to Douglas.

But Douglas got caught by the problems of delivering very sophisticated parts. There were delays and difficulties all along the line and with engines as well, and pretty soon the Pert study stopped matching the cash flow.

The DC-9 turned out to be a winner of an aircraft, timely in respect of economics, the marketplace, and operational technology, but suddenly the bankers were breathing down the corporate neck. Douglas began to look like a high-cost manufacturer—and though this could have been remedied with time, money, and people, Douglas wasn't allowed any of these.

In the end, Douglas was caught by the cyclical aspects of the commercial aircraft business, just as McDonnell, with whom it merged, might have soon been caught by the cyclical aspects of the military and space business if it had not acquired Douglas. Perhaps the merger was a marriage made in heaven, but these days the causes and effects of both cycles are about to change.

So far, I have been discussing the kind of troubles technical inbreeding can bring upon an industry. But these days an industry as large as the aircraft industry simply cannot fall into ill health without having effects on the national economy almost as dire as a setback in steel or in the auto industry. And the air transport industry is a crucial element in the mobility upon which the nation thrives.

If the entire resources of such an industry are marshaled toward the preservation of existing status and capabilities, irrespective of the viability of the end product, there can be widespread economic and social consequences.

The introduction of radically new technology is always upsetting. But it is even more upsetting when there is a failure to perceive critical problems in the timing of that introduction or in the interaction of the new technology with the operating environment.

There is a critical problem here—one which has characterized the aircraft industry and many others that rely on an umbilical cord to the Department of Defense. The builders of a new military aircraft have no hesitancy about investing in a new technology, almost independent of the cost. The criteria for success of their products are vastly different in battle than in the civilian marketplace.

On the other hand, the designers of a product to be sold in the open market have to balance many economic factors: choice of the customer, costs of manufacture, availability of raw material, and the whole problem of marketing.

These determinants soon will apply strongly to many of our defense-oriented industries, for we are coming to the end of many of the large, federally-sponsored, system-management programs which have nurtured them. The nation could not have done without these contractors in developing major systems for defense and for space exploration. But those national requirements are coming to an end.

What can the management of such industries do in this situation?

It seems to me that the aircraft industry does have a massive capability for problem solving on a systems scale and through clever application of high technology. My concern is that such capability has been directed by a perception of what it is *possible* to build rather than of what *needs* to be built. It is time for a reevaluation of the objectives of R&D and the use of production facilities. It is time for a reassessment of the corporate structure itself. It is time for the aircraft industry to reevaluate its connection with the Department of Defense.

It would seem obvious that system management skills and the ability to apply high technology could be useful in many other areas beside aerospace manufacturing. Unfortunately recent attempts by the aerospace industry to convert such skills to urban technology or to medical engineering have revealed evidence of the same problems of technical inbreeding. I suggest that there are more opportunities and tremendous needs right in the field of transportation.

With too few exceptions, the aircraft industry still remains committed to airframe thinking. It is not yet examining enough fresh opportunities for the application of technology to the full spectrum of *transportation* needs   on, over, or beneath land and sea.

For example, there is fantastic need for reassessment of the rail distribution system. There is need to design roadbeds which would improve the speed capability of surface rail systems. Even the most casual observer can perceive opportunity for creative designs to replace that conglomeration of square-wheeled boxes laughingly called rolling stock.

The potential of immediate ocean subsurface transport capabilities— flexible cylindrical containers of the best hydrodynamic shape, towed at high speed by tugs, has hardly been explored. These could transport by sea anything presently moved in a pipeline.

Besides building bigger and better aircraft, there are all kinds of problems that the aircraft industry has the ability to handle, problems that can keep R&D challenged and the production lines flowing and the marketing departments busy. But first the industry must learn to think of itself as a problem solver, not just an aircraft producer.

Simply building an airplane bigger can produce a fearsome economic beast.

Simply building an airplane better can supersaturate a production organization.

If the aircraft industry thinks that just doing more of what it has been doing successfully in the past, without managing the problems this creates, that elephant in the barn might suddenly starve to death.

May 1969

3

The Splintering of the
Solid-State Electronics Industry

One of the most striking features of high-technology companies in the past twenty years is the spinoff. Indeed, around certain small geographical centers, the process of creating new companies—from financing, to organizing people and material resources, to making and distributing products—has become a way of life. Take the semiconductor industry. In the last two years, twelve new semiconductor firms were launched around San Francisco. Most of them, either directly or indirectly, were spinoffs of Fairchild Semiconductor, which has already had a rich spinoff history.

Nilo Lindgren asks whether we may not be seeing the first signs of a fundamental change in the way business is conducted in this country.

Mr. Lindgren is a Senior Editor of *Innovation* Magazine.

From the traditional point of view as to how new industries are formed and evolve, the semiconductor industry, now barely 20 years old, presents some puzzling problems to the industrial anthropologist.

It grows fast, going through generation after generation technologically. Right now, like the computer industry which is its technological peer and natural ally, it is embarked on the realization of fourth-generation concepts. And it shows no signs of losing momentum. Indeed, one of the striking characteristics of this industry is the rate at which it spins off new small companies, and the surprising number of these that survive and grow. Despite predictions that the giants will absorb the little companies and grow more dominant (as the automotive companies did in their earlier stages of evolution) there are signs that an opposite tendency is operative.

In the past two years for instance, 12 new semi-conductor firms have been started in the San Francisco Bay area alone, all of them within a few miles of one another, and there are hints that other new companies are in the planning stage. This is not even counting a half dozen new semiconductor production equipment firms established during the same period.

Our correspondent on the West Coast, Marion Lewenstein, whom we asked to dig out information on these new companies, reports that even in that industry, the people-in-the-know are stunned at the number of new entries into what has been from the very beginning one of the most phenomenal industries our industrially based country has seen.

You might think a saturation point would have been reached long ago. But counting the new arrivals, there are now nearly 25 semiconductor firms all in the same tight area just south of San Francisco. Why the sudden explosion of new firms? Why all in the same region?

As one looks into it, one finds that almost all of these new firms are (directly or indirectly) spinoffs of Fairchild Semiconductor. Fairchild's own Route 128. That in itself is intriguing. Motorola in Phoenix and Texas Instruments in Dallas haven't experienced a spinoff history like that. And one successful spinoff from TI—Siliconix—didn't stay in Dallas. It nestled right up against the others within cannon shot of Papa Fairchild.

One can say, of course, that the attraction of living in Bay area California is part of the answer. It is, but only part.

One can say advancing technology is part of the answer, but that isn't all either.

One could say that what we are seeing is merely a short-term phenomenon, soon to be redressed and normalized, but this "explanation" short-circuits certain other suggestive alternatives.

We might ask, for instance, whether or not the electronics business, as typified in the semiconductor industry, is *fundamentally* different from other businesses? More generally, we can ask whether or not *all* modern high-technology business is *fundamentally* different from the business of the past.

In his book *The Limits of American Capitalism*, the economist Robert L. Heilbroner speculates persuasively that modern technology is a major glacial force changing the basic character of capitalism itself. Is it too far-fetched to imagine what we see in the semiconductor industry is a perceptible instance of such a fundamental change? Rather than living through a short-term event, typified by the trade press mind as "exciting" and "spectacular," might we already be going through the threshold of a more fundamental change whose long-term ramifications and consequences are not yet clear?

There is some evidence to support such a view.

If we are really in the process of such a fundamental change, what does it mean for the manager of advancing technology? How should he "read" his role?

Such speculative questions can hardly be answered conclusively, but they might be kept in mind as a frame of reference for the study of the semiconductor industry of the Bay area. We suggest that Fairchild's own Route 128 might be viewed as a case study of a more general phenomenon, the nature of which should be actively and vigorously explored.

The immediate reasons for the phenomenal growth of semiconductor companies near San Francisco are many and complex, but the character of the Fairchild Semiconductor Company, as the new president of the parent company (Fairchild Camera & Instrument) Dr. C. Lester Hogan, points out, plays a major role. The fact is that Fairchild was started by eight extremely bright young men, and they attracted to them a group of highly energetic, bright professionals, both in the technical sense and in the marketing sense. This gave Fairchild, Hogan goes on, more than its share of industry leaders who could potentially strike out on their own, and subsequently did just that.

The result of this extraordinary spinoff history is shown on the family tree. New firms are indicated. Although the family tree shows that some of the new company organizers come immediately from other companies than Fairchild, almost all at some point in their careers worked at Fairchild. The map shows the small region where the dozen new companies are located.

In the same region is Hewlett Packard Associates, a semiconductor manufacturer, which is extraordinary for another reason. It is virtually the only firm in the region with no history of spinoffs—only three individuals have left the company so far as anyone knows. The answer? HPA has always had a strong policy of making their people feel comfortable and necessary to their job. Further, long before it became fashionable, the company had stock option plans that extended far down into the company, to the level of some factory workers. (In the past, Fairchild had no such stock options.) Although HPA is beginning to get big and unwieldly now, and there begins to be dissatisfaction, the image hangs on—"Hewlett Packard people stay."

Although the story of the "original eight" founders of Fairchild Semiconductor has been told many times by people in the electronics industry, the highlights need recapitulation here. In 1955, a half-dozen years after his co-invention of the transistor at Bell Telephone Labs, Dr. William Shockley left the Labs and after a brief flirtation with Raytheon in the East, returned to his native Palo Alto to start his own company, Shockley Transistor Corporation. That, in fact, was the beginning of the semiconductor industry on the San Francisco peninsula. Shockley was successful in attracting many really bright physicists and engineers, but was not so successful in his personal relations with them. This was one of the factors which (less than a year after he won the Nobel prize for his work leading to the transistor) helped lead to the departure of the group of eight who founded Fairchild Semiconductor in Mountain View, California. That was in September 1957 with the backing of the Fairchild Camera & Instrument Corporation whose headquarters are on Long Island. The eight founders were Victor Grinich, John Hoerni, Jay Last, Sheldon Roberts, Eugene Kleiner, Julius Blank, Gordon Moore, and Robert

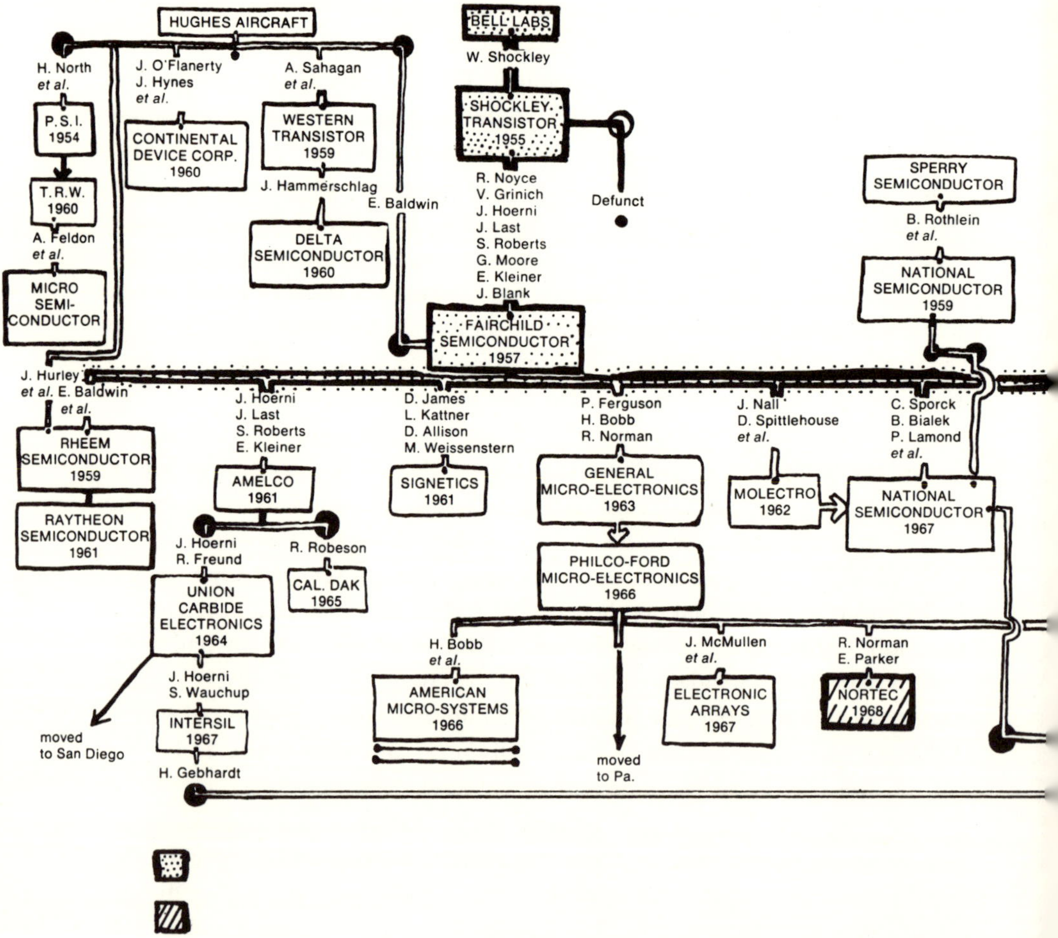

Noyce, who two years later, in 1959, was put in charge of the Fairchild Semiconductor operation, and who managed it over most of the time of its spectacular growth.

Of the eight founders, Hoerni, Last, Roberts, and Kleiner left in 1961 to found Amelco. Subsequently, Hoerni went on to found Union Carbide Electronics (1964) and Intersil (1967). When Amelco was purchased by Teledyne, Last stayed on to become a vice-president; Roberts went on to become a consultant, and Kleiner first founded Edex, a nonsemiconductor firm, and then began dealing in venture capital for electronic firms. Noyce and Moore left Fairchild in mid-1968 to form their own firm, Intel. Blank, the last to leave, joined Ness Industries, management and venture capital consultants in high-technology fields. Grinich, who had been teaching, rejoined Hoerni as

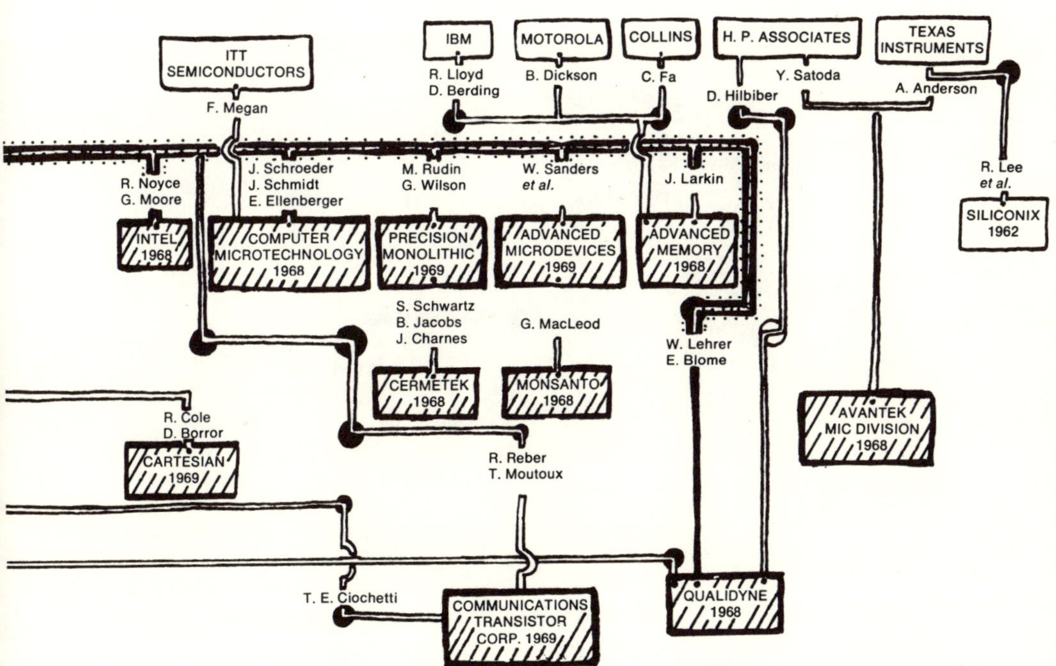

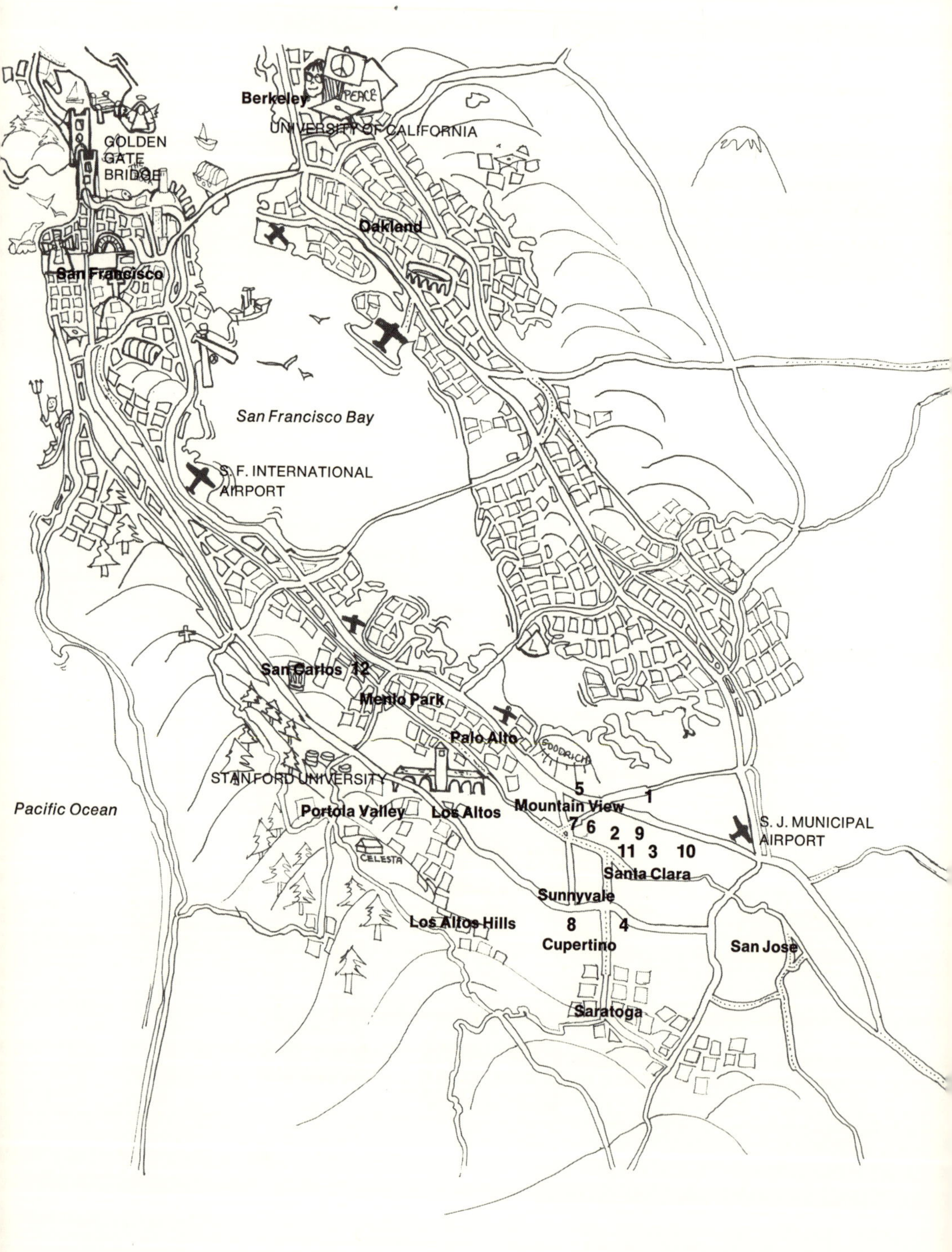
Berkeley
UNIVERSITY OF CALIFORNIA
PEACE
GOLDEN GATE BRIDGE
San Francisco
Oakland
San Francisco Bay
S. F. INTERNATIONAL AIRPORT
Pacific Ocean
San Carlos 12
Menlo Park
Palo Alto
GOODRICH
STANFORD UNIVERSITY
CELESTA
Portola Valley
Los Altos
Mountain View
5
1
7 6
2 9
11 3 10
Santa Clara
Sunnyvale
S. J. MUNICIPAL AIRPORT
Los Altos Hills
8
4
Cupertino
San Jose
Saratoga

an advisor to Intersil. One might note that the eight have alone been directly responsible for the formation of at least five new companies since Fairchild, although only three men—Noyce, Moore, and Hoerni—are still directly involved in semiconductor manufacturing. The original Shockley Transistor Corp. meanwhile was bought by Clevite in 1959, then by ITT in 1965. ITT moved the R&D work to Palm Beach, Florida, in 1968, and shut down Shockley's Palo Alto facility in 1969. Shockley himself moves between Stanford University, Bell Labs, and other non-semiconductor activities.

Though Fairchild grew fast after the breakaway from Shockley in 1957, it experienced its own first spinoff in 1959 when E. M. Baldwin left to form Rheem Semiconductor; and then in 1961, four of the original eight—Hoerni, Last, Roberts, Kleiner—formed Amelco. This was quickly followed by the formation of Signetics also in 1961, of Molectro in 1962, and of General Micro-Electronics in 1963, all out of Fairchild. Of the original eight, Hoerni must surely be the most restless or driven—since forming Amelco in 1961, he has gone on to form Union Carbide Electronics in 1964 and Intersil in 1967.

In retrospect, considering these and other subsequent spinoffs, one could conclude that the semiconductor industry (besides being what it is—one of the most competitive, gut-eating industries going) has also been one of the most unusual and spectacular schools of technical management in this country. It should be remembered that Shockley and the men he attracted were scientists, trained at a time when scientists expected to be just that, and before the tremendous financial rewards of semiconductor theory had become overpoweringly evident.

When Noyce, for instance, was made manager of Fairchild's operation, he was only 32, he had no management experience, and more than that, some reluctance to become one. He resisted going with the defecting group at first, though he was prevailed upon. And when he was offered a management position, he took it partly because of an incident that had occurred during his work with Shockley. During the clashes that had gone on there, Shockley had conducted a straw vote among his thirty or so colleagues, asking who among them they believed would be the best technical leader. Everyone voted for Noyce! That expression of confidence persuaded Noyce, when the time came, that he should step into the position of making decisions and being responsible for them.

When Noyce and Moore left Fairchild in July 1968 to form a new company, followed not much later by the last of the original eight, Julius Blank who joined Ness Industries, it marked, in the world of electronics, the end of an epoch.

Noyce's departure was duly observed by the investment world. Fairchild stock took a sharp dip, the electronics and financial communities spawned rumors faster than the production lines turned out new semiconductor circuits, and Fairchild lived for the space of a breath without a leader. Then, C. Lester Hogan was lured from Motorola by Sherman Fairchild to become the new president of Fairchild Camera & Instrument Corporation, and hot after him came some of his erstwhile Motorola colleagues and an enormous lawsuit which is still to be settled.

It was not long before Hogan made his presence felt at Fairchild. From the outside, there was clearly, as the stories of departures and new appointments made the news, a lot of reorganizing going on as Hogan reshaped the company. From the inside, as Hogan studied the internal economics of the Fairchild operation, a fascinating insight emerged. Hogan discovered, in the retrospective of the records that in May 1965 there was an abrupt change for the worse in all the internal records—in sales, production, and so on. That sharp break came in exactly the same week that Noyce had left active management to act as a vice president of the parent company in the East! It showed, in Hogan's view, how brilliantly strong Noyce's command was of the semiconductor operations. Both Hogan and Noyce have concurred in stating that certain important changes were needed in Fairchild about that time. Hogan testified, "Bob Noyce made my job easy for me. He had already done all the hard things that needed to be done in building Fairchild, and he left the easier part for me."

Among other things, in fact, Hogan was able to attract some top quality semiconductor professionals who had not previously been at Fairchild. From Hewlett Packard, he got M. M. John Atalla; from Bell Labs came James Early; from Stanford came John Moll, a full professor.

Meanwhile Bob Noyce and Gordon Moore had gone on to form their new company, Intel, as had been rumored, a company that almost everyone predicts will be as successful as was the original Fairchild. With its first products just coming into the market ("right on schedule"), Intel projects at least $25 million in sales within the first five years. Some people think they might do twice that. Noyce believes they will build up a $100-million business "over a period of time."

Moore, Intel's vice president, stresses that although there is lots of room for specialization, and therefore room for new firms, the only way for a semiconductor company to stay in the mainstream is to get big. Yet, with the change at the top of Fairchild, which all (including Noyce) agree is for the best, the parent company is expected to remain a major and alert contender in the semiconductor business. Can there be two companies of the same great size in the same business? Much less 12? What are the chances of these new spinoffs?

The early rash of Fairchild spinoffs, in 1961-1963, arose partially because many key people (including four of the original founders) felt frustrated by the increasing bigness of the company and by the refusal of the conservative Eastern headquarters to give stock options or profit-sharing. One of Hogan's first acts, in fact, was to pass along a greater share of the action down through the lines with stock options. But the new spinoffs do not seem to come from dissatisfaction. Rather the reverse. The evolving technology of semiconductors is once again opening up handsome new opportunities.

Generally speaking, the spurts of real growth in microelectronic firms have come with breakthroughs in technology. For instance, Texas Instruments succeeded with its initiation of silicon transistors, Fairchild with the so-called double diffused and planar techniques, and Siliconix with so-called junction field effect transistors (FET's).

What is happening today in the semiconductor technology is potentially of a different order. As one looks at the new companies emerging, one could break down their product lines into two rather traditional categories—those that are going after different specialized segments of the market, and those that are exploiting the forefront technology to go after the colossal computer market. The latter category includes the so-called MOS (metal-oxide-semiconductor) and LSI (large-scale integration) technologies. The name LSI refers to the semiconductor chips that hold an enormous number of circuits rather than just a few components. For instance, one of the first products being developed by Intel is an entire 256-bit memory that goes on a chip 1/10th of an inch on a side! A great deal is expected in the field of advanced semiconductor memories for computers, especially if computer manufacturers do some redesigning to take advantage of such circuits.

What this means is that as computers become larger and faster, the standard concepts of their architecture are changing. It is said that the use of small, fast, more expensive semiconductor memories to act as buffers to assist the core memory could increase the performance of existing core memories by as much as 300%. More attractive, but perhaps further away (as volume production drops the cost), is the use of semiconductor components for the core memories themselves. Thus, there is a push in the new small companies to perfect semiconductor memories and to persuade computer manufacturers of their advantages.

Another enormous new market growing out of the advancing technology will be for light-emitting devices, which the experts believe will find many possible uses (in digital readout, and so on).

But there is an aspect of the advancing semiconductor technology that goes even beyond these traditional directions, as vast as they are. The plain fact is, hardly anyone (in all the other industries outside the semiconductor field) really grasps the enormous amount of electrical circuitry he now has available to him "on the head of a pin." That is, LSI's, or even smaller units, can undertake tasks in industries that never before used any electronics—all kinds of industrial machinery, for instance, could be redesigned for greater effectiveness and controllability through the incorporation of what amounts to the "intelligence" functions of LSI. Automobiles will certainly use more such microminiaturized functions—small cheap computers for safer driving, and the like.

This penetration of the semiconductor technology into other industries is really just beginning. It is going to require much "custom" designing of integrated circuits, and the redesigning of industrial (and home) equipment, but the semiconductor industry is consciously mobilizing itself for precisely this end. The consequence of this twin thrust—of semiconductors and computers—is that the Second Industrial Revolution (that of integrated information functions in all machines) that Norbert Wiener predicted 20 years ago is fast upon us (maybe in light of recent events, we should call it the Age of Electro-Aquarius).

In any event, much of the explosive spinoff and reorganizing within the semiconductor industry may be regarded as partly symptomatic of a broadening of aims allied to a consciousness of specific new markets.

But what do the new company founders say about themselves? Let's take a quick look in the next pages.

> **Semiconductor Country: an Overview:** Within the San Francisco Peninsula area, semiconductor firms are moving southward. Palo Alto, the initial location for Shockley and early Fairchild, is 35 miles south of San Francisco. Twenty five or more semiconductor facilities are located in this one small area. The furthest from one another is probably only 17 miles . . . 20 at the most. Many are only ¼ mile off the extensive and swift freeway system, the farthest from a freeway is about 1½ miles. Airport connections, also important to the semiconductor industry, are excellent. San Jose has a growing airport; San Francisco has worldwide service; and fast helicopter service links up the major points of the whole Bay area.
>
> As the population increases in this area, all the newcomers who moved from crowded Eastern seaboard locations see some of the same problems of transportation, smog, hustle-bustle they thought they had left behind. So the impulse is to move down toward San Jose where there is more open, cheaper land and fewer people. Some communities—like Sunnyvale's Science Industry Park and Santa Clara's Space Park—are trying to attract more of the electronics industry. The executives of the many semiconductor companies live mostly in the plushier areas west of Palo Alto—the Los Altos Hills, Saratoga, Portola Valley—where it is easy to commute to Santa Clara, Sunnyvale, Mountain View.

1. Advanced Memory Systems, Inc., is aiming mainly at developing so-called bipolar memory semiconductors, the technology in which the founders have had the greatest combined experience. This area, the founders feel, is not being properly served now. Also, there is less existing competition in the system area, so these people are using designers with computer experience to come up with customer-oriented products. The staff held middle management positions in advanced technology in their previous firms.

Founders: Robert Lloyd, president—formerly with IBM, manager of memory device development; Jerome Larkin, vp marketing—formerly with Fairchild, IC product manager; Brent Dickson, vp manufacturing—formerly Motorola, manager LSI production; Drew Berding, vp engineering—formerly IBM, manager high-speed memories; Charles Fa, vp technology—formerly of Collins, Newport Beach, manager IC technology.

Capitalization: private plus two mutual funds.

2. Advanced Micro-Devices, Inc., whose entire group of founders came from Fairchild, is considered to be most "market oriented" of all the new startups. The firm will specialize in medium-scale integrated circuits (MSI) (both linear and digital types), and will emphasize its processing capability to distinguish itself from the big firms who can offer low costs through mass production. MSI complex linear and digital circuits, which are difficult to build, depend more on the kind of customized engineering that a small company can provide.

Founders (all formerly of Fairchild): Walter J. Sanders, president; Jack Gifford, vp marketing and business development; John Carey, vp complex digital integrated circuits; Larry Stenger, vp complex linear circuits; Edwin Turney, vp sales & administration; Jim Giles, director linear circuit engineering; Frank Botte, director linear circuit operations; Sven Simonsen, director complex digital engineering.

Capitalization: investment houses and banks from both Coasts and Europe.

3. Avantek, Inc., which was started in 1965 for production of broadband microwave components in microelectronic configurations, has launched its Micro Integrated Circuit Division to fulfill the need for highly specialized, exotic devices that the large companies are pretty much ignoring. Thus, by satisfying its own need for high-frequency, high-gain, low-noise microwave transistors, amplifiers and oscillators, and by keeping pace with developments in other microwave areas, the company can take advantage of a growing advanced equipment market of other users as well. Because it had no semiconductor background, Avantek lured Dr. Yozo Satoda from Hewlett Packard Associates and Dr. Andrew Anderson from Texas Instruments to head up the MIC division.

Founders: Lawrence Thielen, president—formerly of Applied

Technology and Ampex; Dr. Yozo Satoda, manager MIC division—formerly of Hewlett Packard Associates; Dr. Andrew Anderson, manager active devices—formerly of Texas Instruments.

Capitalization: Avantek (founded 4 years earlier) is parent firm.

4. Cartesian, Inc., which was phased out of a mask-making company affiliated with Electromask, Inc., will engage in circuit wafer fabrication and LSI masking for those other firms that can design and test their own circuits. What Cartesian offers is lower cost for mass production of LSI devices through low overhead and without high-priced circuit designers. Part of Cartesian's "talent" philosophy is to look not for supermen, but to get competent middle management from older firms where such men have no access to stock options.

Founders: Gerald M. Henriksen, chairman of the board and current president of Electromask, Inc., Robert Cole, president—formerly of Philco-Ford General Micro-Electronics, earlier with Fairchild—then with Cartesian Corp., an affiliate of Electromask, Inc., Daniel R. Borror, vp—formerly of GME, before that Fairchild, then on to Cartesian Corp.

Capitalization: investment banking.

5. Cermetek, Inc., will concentrate its products on high-speed, high-voltage hybrid devices with high reliability, a field in which designer-founder is said to have no competitors. For instance, the firm now has a high-speed clock driver for MOS circuits that is said to be unique. The company expects to keep a lead in mass production of hybrid MOS devices (nearest competitors are Signetics and National Semiconductor) through a licensing arrangement whereby Components, Inc. (located in Maine and in Phoenix) will produce Cermetek designs, a mode of operation that Fairchild and other firms practiced in the past.

Founders: Samuel A. Schwartz, president—formerly design consultant to Fairchild, ITT, and others—also research scientist with Lockheed Missiles and Space; James Charnes, executive vp—formerly with components division of Sprague Electric, before that with Burroughs Corp.; Dr. Bernard Jacobs—formerly of General Instruments, now resigned from Cermetek.

Capitalization: privately held, and through investment firms; probably go public in year or two.

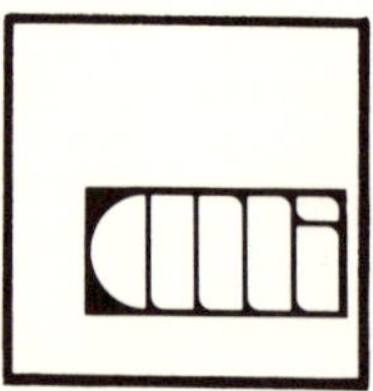

6. Computer Microtechnology, Inc., will aim at the area that is clearly going to be the biggest future market and that is also going to be most competitive—computer memory devices. In a big and growing market, president F. J. Megan argues that there will be no single source and that all companies will get their share. Those who have a good grasp of the memory technology and who are (as the currently popular phrase describes them) "people sensitive," believe that it "will be hard not to be a success" despite the competition.

Founders: Francis J. Megan, president—formerly of ITT, Florida; John Schroeder, vp process development—formerly of Fairchild; John Schmidt, vp engineering—formerly of Fairchild; Charles Ellenberger, vp management—formerly of Fairchild.

Capitalization: personal from founders, some seed money from other personal investors and investment sources.

7. Intel, Inc., started by Noyce and Moore from Fairchild, will push hard with LSI varieties of advanced memories. These men expect that by concentrating on memories, these products will get cheaper and better, and that Intel will keep a lead on the giants like Fairchild, Motorola, and TI, whose efforts are more dispersed. Although Intel is expected to grow fast, it is significant that Noyce was partly motivated to leave Fairchild because he prefers keeping closer touch with the laboratory research, which is easier in a small company atmosphere. He sees the small company as allowing more "human involvement," as well as providing the chance to make big money. Noyce and Moore both indicate they were getting stale in their old jobs, and uncomfortable as the situation at Fairchild changed, so that they felt a loss of loyalty from both above and below. Now, with Intel, these men can explore a major new market opportunity with the decision-making power in their own hands.

Founders: Dr. Robert Noyce, president—formerly of Fairchild; Dr. Gordon Moore, vp—formerly of Fairchild.

Capitalization: personal investment, principally by founders but otherwise from a few other people in the firm and other private sources.

8. Monsanto Co., headquartered in St. Louis, which has a background in developing materials used in photo emitting devices, is now setting up an off-shoot called Electronic Special Products. The company will concentrate its product line on photo emitting or photo-optic electronic devices. Not only is this product new but so are its applications. However, the market projection is $300 to $500 million/year within ten years, and Monsanto is a strong bidder for a healthy share. Like other established companies who have come to the San Francisco Bay area, Monsanto has the same reason—this is where all the high-technology people are, where new developments are assimilated virtually by osmosis at cocktail parties and the like. George M. MacLeod, general manager of the new operation, as a big company representative must struggle with a conservative Eastern management to supply the motivation other small new companies can offer top professionals (i.e., stock options). But he has an alternative in his deck: He is setting up each area of research in light emitting devices as a miniature business setup, with its own R&D, its own marketing, and so on.

Founders: New division of Monsanto, George M. MacLeod is general manager.

Capitalization: parent company.

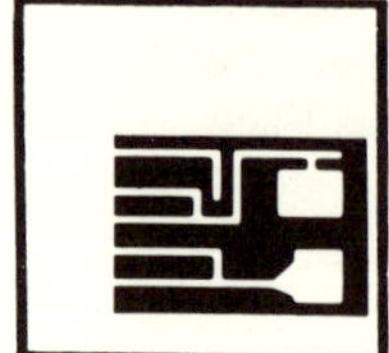

9. Nortec Electronics, Inc., whose president Robert H. Norman was one of the founders of General Micro-Electronics in 1963, and who was with Fairchild before that, will specialize as fabricators in MOS and LSI production. Rather than compete with Fairchild, Signetics, Amelco, and others, the company will use its customers own designs rather than generating its own proprietary circuits. With the mask provided by the customer's engineers, Nortec will make the wafer, dice, or package. In practice, this differs from Fairchild custom work, for example, in that Fairchild works out the complete configuration and then fabricates it. The customer must reveal proprietary information about his end product and then pays a markup on engineering time as well. With the ever-widening use of semiconductor circuits, there has been an increasing need for semiconductor fabricators who will work to a customer's mask, and there are, as well, an increasing number of companies that have staff engineers who can design circuits for MOS or LSI production. (Hewlett Packard, incidentally, started its internal semiconductor operation to protect its proprietary information. At first it sold most of its production within the company, but has progressively increased external sales, and now is one of the top five semiconductor producers in the San Francisco area.)

Founders: Robert H. Norman, president—formerly of Fairchild, founder of GME, left there for Nortec after GME was purchased by Philco; Thomas L. Turnbull, vp financing, formerly of Applied Technology; Edgar R. Parker, vp operations, formerly of GME, stayed on after Philco purchase but went to Fairchild when GME moved east.

Capitalization: individual investments, including founders.

10. Precision Monolithic, Inc., aims at moving fast with an initial big financial investment in order to "get to the top of the heap." Its product line, to start, consists of unusually precise analog integrated circuits for special processors and for peripheral computer equipments and special circuits for digital-to-analog converters. The founders reason that by putting together an electric capability by recruiting people from instrument, aerospace, and circuit design firms, and by bringing in people with good judgment who have not been "isolated in top management," they will help create a healthy company, where the research teams will be small and where each individual will see the results of his contributions. One feature of Precision's backing is that Bourns, Inc., an electronics components company in Riverside, California, has put up 40% of the initial investment with the option to acquire controlling interest of the firm by mid 1974.

Founders: Marvin Rudin, president—formerly of Fairchild; Dr. Garth H. Wilson, vp—formerly of Fairchild.

Capitalization: 40% by Bourns, Inc., an electronics components company from Riverside, Cal.—60% private, mostly from within company.

11. Qualidyne Corp., whose present president, H. Ward Gebhardt, was previously a founder of Intersil (in 1967), takes a tack contrary to all the other new startups. Its founders believe that it is not necessary to start out with a unique position, either technically in its products or in its potential markets. New items, they point out, are not company money-makers. For instance, Intersil, which never bothered becoming a technological leader, nonetheless goes on earning money. So, Qualidyne says, it will reverse the usual procedure in becoming a leader—it will do its technological development *after* it is established. This it will do by producing custom and standard sense amplifier integrated circuits for computer core memories. Most of the firm's business thus far is as a "second source" producer for devices pioneered by other companies. For the future, the company feels it has up its sleeve the capability of marrying micro-resistor thin-film techniques with linear integrated circuits, which so far no one else has been able to do successfully.

Founders: H. Ward Gebhardt, president—formerly of Fairchild and among founders of Intersil; Dave Hilbiber, former president, now left—formerly of HPA and Fairchild; William Lehrer, manager thin film—formerly of Fairchild; Eugene Blome, manager photo masks—formerly of Fairchild.

Capitalization: venture capital and investment of founders.

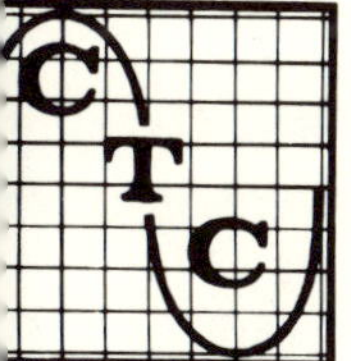

12. Communications Transistor Corp., the newest of the new dozen, formed October 17, 1969, also has three founders who have been at Fairchild, although its president, Thomas E. Ciochetti, was most recently with National Semiconductor. The new company will be located in plant space at Varian Associates' Eimac Division and will be a Varian affiliate. It will produce high-frequency transistors in the UHF, VHF, and microwave range, approximately the same range served by conventional Eimac klystron tubes. Thus, it seems clear that CTC is planted cheek to jowl with its first "natural" market. Varian holds equity in the firm, and some Varian officers will sit on the CTC board. However, the officers of both companies insist that CTC will preserve is own distinct identity.

Founders: Thomas E. Ciochetti, president—formerly of National Semiconductor and Fairchild; Robert Reber, vp—formerly of Fairchild; Thomas Moutoux, vp engineering—formerly of Fairchild.

Capitalization: affiliate of Varian.

That's the quick rundown on the rough technological positions and philosophies of the new dozen. What are their chances for success?

Knowledgeable insiders in the semiconductor industry say that Intel (7) will "absolutely" make it, for not only are Noyce and Moore technical geniuses, but they have their already spectacular track records at operating a company. Cartesian (4) has a smart formula, but there is some doubt about how large it can grow. Monsanto (8) seems a good bet on the basis of its unusual technology. Avantek (3) may become leaders in a specialized area and financially lucrative, but not likely to become giants. Advanced Memory Systems (1), Cermetek (5), Computer Microtechnology (6) are all given a "reasonable chance" on the basis that they have the ingredients for success. Precision Monolithic (10) is thought to have smart enough leadership, though the firm may not be specialized enough. Nortec (9) and Qualidyne (11) are looked at dubiously mostly because they lack high-technology capability, and it is pointed out that Qualidyne's "reverse" formula has not previously succeeded in the semiconductor industry. Most uncertainty seems to register on Advanced Micro-Devices (2).

That is, of course, only one set of starting odds on the new dozen, and they leave out certain interesting possibilities. For instance, the general feeling in the semiconductor community in the West is that the market is so bullish for the evolving MOS technology, the diversification of products in light of ever-widening applications creating room for all, and the semiconductor memory business so staggering in its potential, that it will be hard for anyone with any competence to lose. Even if all these firms are not around five years from now, the founders, it is said, will have made money by selling out to others. Certainly, the new firms with big parent backing (who hold options to buy controlling interest later) fuel this image. The Westerners think of Raytheon's acquisition of Rheem and Philco-Ford's acquisition of General Micro-Electronics.

So, it is said, no one will lose, for all these people are competent. As one founder said, "We all know each other, and we all respect each other's capabilities." But only a few, it is assumed, will really make it big. And no matter what they all say about their motives—the challenge of going after new markets, the ego satisfaction in being top dog, the excitement of risk-taking where the potential for reward is good, the pleasures of a small company—there is little doubt that the prime motive for most is the potential financial reward.

However, the situation is also read more darkly by some observers. They feel that some of these new companies are being deliberately set up for luring in public money, and that the founders will sell out, enriching themselves and leaving others to hold the bag. Said one long-time industry observer, "Most of these companies are parasites, the modern day equivalent of gold-mining stocks." Little companies, he says, do have a real economic function in developing markets for a radically new technology like holography. But when a technology is mature, he concludes, as it is with semiconductors, there is a real question as to whether the little company is performing a socially useful function.

Why the semiconductor industry should have grown so strongly on the San Francisco Peninsula has been partially answered and has been partially self-evident.

The obvious reasons that everyone cites, once Fairchild's germinal role is acknowledged, is that besides being a pool of major talent, it is the best region for recruiting lower levels of personnel such as senior technicians, laboratory workers, assembly workers (mostly girls), and so on, who would be expensive to move across the country the way engineers are. The availability of support people with 5 to 10 years' experience in such work as machine tooling for dies, saves new companies lots of training time. Moreover, it is easier to lure such workers to change jobs because they know they can always get another job in the same locale if the new company fails.

On top of this is the availability of materials and equipment—vendors of vacuum equipment, silicon and industrial gas (clean nitrogen, for instance, is hard to get in regions where winters are snowy). Because the vendors are familiar with the problems of the semiconductor customers, they are able to offer better services.

Also, Richard Lee, president of Siliconix, notes the influence of the University of California, Berkeley, and Stanford. Between the two, they are probably doing the leading solid-state work in the U. S., Berkeley being strongest in solid-state applications while Stanford is strongest in original development.

On that point, Robert Noyce points out that it was after the original success of Fairchild that the two schools became important supporters of the technology. If the Shockley-Fairchild ventures had not brought an influx of people interested in the field, Noyce believes the two schools may not have become the solid-state leaders they now are.

Not so obvious is the existence of a strong financial community that has confidence in the semiconductor industry. Because of the existence of past winners, there has been bred a sense of confidence among investors as well as among professional employees.

Still less obvious to the outsider is the atmosphere of mutual help that pervades the peninsula and that is beneficial to the new company. One new company founder (James Charnles of Cermetek) said, "If I need something, I know I can walk over to Fairchild and get help." Unless one is directly competitive with another firm, one can get aid even from a relatively competitive firm.

Everyone in fact speaks of the enormous camaraderie in the region. Despite their being competitive on a business basis, and their business willingness "to wipe someone out," the high-technology people mix frequently socially, exchange ideas, and have a genuine high regard for each other. Most of the top management people from different companies are social friends. Said one, "It is not like GE and Westinghouse top management who are not likely to become close personal friends."

One fact certainly stands out. Most of the people are ex-Fairchilders, but not even those who were forced to leave Fairchild recently seem to have anything nasty to say about Fairchild or Lester Hogan.

Hogan himself has been struck by this since his arrival, and he was unprepared for it. He sees a number of reasons behind it. For one thing, over the years of working together, many friendships have developed. When you see people, even rivals, in relaxed social conditions, Hogan says, you learn to like their human qualities. You learn to assess their strengths and weaknesses, and you know in what areas they will likely succeed and likely fail.

Another factor is the geographical closeness, with many industry people living and working nearby, so that encounters are frequent. It's hard to build up images of "the enemy," Hogan remarks, when you see someone often. In contrast, when you are isolated, as you would be in Phoenix or Dallas, for example, without frequent contact with your competitors, it is much easier to build false images.

There is, too, a family feeling among semiconductor people, and Fairchild is the father. That's why, Hogan says, they don't speak harshly of Fairchild—it would be like attacking their father. This is confirmed even by some new-firm founders who were displaced by Hogan. If they were to continue working for a salary, they say, they would prefer working at Fairchild rather than any other firm because the parent company, with its big resources, allows an individual more flexibility in research and a chance to be creative.

Why then don't they go back to Fairchild when the new venture sours, say, through personality conflicts? Says one man who has started more than one firm: "A certain momentum builds up. When you have started one successful firm, it is difficult to go looking for a job. There is a tendency to regroup with other people and start yet another firm." Pride is certainly an ingredient—a man with enough ego to launch his own company is not going to find it easy to admit failure. Another reason, of course, is that even stock options in another company are not as attractive when you get a taste of the potential financial rewards by starting your own.

One must also look at the other side of the picture: Why didn't new industries spring up around Motorola and Texas Instruments? One reason given is that Phoenix and Dallas are not really as attractive places to live, the climate is too hot, and so on.

A case in point is Siliconix, Inc., whose principal founders, all from TI, chose to emigrate to the Bay area rather than stay in Dallas, both because they personally preferred it and because they knew it would be easier to get the high-technology people they wanted to move from the East or Midwest rather than the reverse.

Whether climate and the metropolitan attractions of San Francisco were the prime movers or not, the fact remains that Motorola and TI never could attract the number of high-calibre, high-technology people that Fairchild did. In the semiconductor community, those companies were considered too mundane and were unable to attract many "genius personalities." But Fairchild started off with so many of the bright, creative types who had the urge to do it on their own that they started a chain reaction that now reaches down even to mediocre people. "If so-and-so can do it, so can I," is virtually a slogan of the Bay area companies. The aura of Fairchild accomplishment hangs on.

But there are gnats in semiconductor paradise too. Some professionals are sometimes reluctant to move West because the Bay area has a high cost of living, is smoggier and more crowded than other regions, and has a reputation for more hippies, more drugs, more unrest among the young. They do not wish to expose their children to "immoral" California. But this reluctance is exhibited virtually always from Midwesterners, not from those on the East Coast.

A persuasive aura of well-being comes through all the descriptions of the Californians. It is almost enough to make an Easterner bolt for the semiconductor paradise, especially as winter dulls the skyline of New York and the icy slush impedes progress along the sidewalks. But then, New York is the publishing capital of the U. S. gathered tightly on the small island of Manhattan. Everyone knows everyone, and even competitors meet often socially. There are publishing spinoffs, and new magazines like *Innovation* are started. There is an esprit de corps and . . . well . . . a common sense of superiority. The real "professionals" in any field, one supposes, are the ones who enjoy doing what they are doing.

And what they are all doing together is an open-ended proposition. An evolutionary (or perhaps revolutionary) process has been set in motion, in which all groups in society seek participation, and the outcome of which is yet to be crystallized. For the short run, it is certainly not stability.

December 1969

4

The Battle for Data Communication

It is a distinctly modern struggle, founded on technology, animated by change, moderated by government regulation. More than that, it pits the telephone company against the computer industry—a sober, conservative pillar of American economy against an adventurous, often brash young giant. Charles J. Lynch examines the issues from both sides.

Mr. Lynch is a Senior Editor of *Innovation* Magazine.

In the early 1960s, scarcely anyone talked about computers communicating with one another, and the term "computer utility" had yet to be invented. But such talk as was heard left very little question about who would benefit from the data communications business if it ever got going. It was a foregone conclusion that whatever business there was to be had would somehow split between AT&T and IBM. The only question was precisely how these two giants would agree to divide it up.

It is now a decade later and neither IBM nor any other computer manufacturer is taking any direct part in the data communications business. Whatever the reason for their disappearance they have left AT&T—unsinkable queen of the seas, veteran of a thousand battles and a million crossings, unequaled giant of American industry—sailing majestically into uncharted waters. Which is not to say that she is alone. Already there are periscopes on the horizon. Already there are gunboats at the flanks. And could that be the sound of water leaking into a forward compartment?

Not, I should add, that it looks this way from the AT&T executive offices. Those periscopes out there have always been there, young man, and those gunboats alongside are probably just innocent fishing trawlers. As far as the uncharted seas are concerned, the seas ahead are always uncharted. But we have the finest navigation equipment that money can buy, and the finest navigators, and we've worked this out to several decimal places, and where do you suppose the boy is with the silver tea service?

A first-order explanation for those menacing shapes out there in the darkness is relatively simple. The transportation of data from one machine to another is a rapidly growing business and, by some counts, promises to equal the voice telephone business—now worth about $14 billion per year—by the end of this decade. According to AT&T, the new business is rightfully theirs and can easily be accommodated on the telephone network, today and in the future. According to the data processing people, AT&T's network was designed for voice and is inherently unsuited to the language of machines. The telephone company should either be required to set up facilities for data transmission, or yield the business to competitive networks especially designed for that purpose.

The real issues are a good deal more complex than that, as we shall soon see.

I am not so brave as to speculate on what the outcome of this struggle will or ought to be. It is a very complicated matter that turns upon a tightly interwoven combination of technology, economics, and government policy. It is worth recognizing, however, that there are certain limits to what will happen. No one is predicting, or even suggesting, the collapse of AT&T's empire. She is too powerful, too well protected, too vital to the nation's welfare for that. It is not even quite a question of whether she will have some of her present business taken away from her, but only whether she will be allowed to exercise all of her potential opportunities for growth.

The answer has a special importance for the managers of technology, not merely because they are users of communication and data processing equipment, but because to an important degree, the change going on in the communications industry is representative of changes occurring or about to occur elsewhere.

It is, first of all, a change that does not hinge upon a single technological development—no jet engine, no transistor, no electron microscope. It is rather a creeping sort of thing that, nevertheless, can turn the whole technology upside down in a matter of a few years. In this case, it issues out of a combination of developments: a decreasing cost of computers, an ever-widening need for computational power, availability of low-cost communication techniques. If there is anything approaching technological suddenness, it is the sudden appearance of what we might call The Time-Sharing Principle— that it is cheaper to use a small piece of a big machine than it is to use a big piece of a small machine.

Secondly, it is happening to the largest system in the world. The telephone system is The System, the cradle of systems engineering, a $45-billion network that represents the largest capital investment of any privately held system in the world. Bringing about change in such a gargantuan enterprise, even modifying it significantly, is an enormous undertaking.

And finally, it is a change whose character and speed will be strongly influenced by government action. Communication is, of course, a regulated activity; data processing is not. Precisely where the new industry falls into the regulatory spectrum is, at present, the subject of a very lively debate. For most people, it is the key issue in the entire controversy.

But before we get into that, let us take a closer look at the cast of characters.

Over here we have what the government calls the "common carriers." Strictly speaking, this refers to anyone who "holds himself out to the public as a regular business to transport property for hire" but in our case it is limited to the people in the telephone and telegraph business. Foremost of these is, of course, AT&T. It is foremost to such a degree that the other carriers—among which are General Telephone and Electronics (the largest non-Bell company), Western Union, United Utilities, and a group of other independents—are left with less than 10% of the action. Beyond being big, AT&T is generally well managed, and has a record of service and innovation that is rare for a regulated public utility of any sort. Its principal weakness in the present circumstance is that it is an old-line company, with all that the name implies.

Over here in the other corner we have the data processing people, represented mainly by the main-frame computer manufacturers, but also including the accessory manufacturers, the softwear producers, the service bureaus, and time-sharing operators. In spite of the fact that most of them are not actual contestants in the battle for the data communications business, they play an essential role. It is they from whom the data communications business comes, and it is they who voice disappointment with the present levels of data communications services. It is they who fan the flame of controversy.

Interestingly enough, they are one of the few adversaries capable of giving the telephone company a match. The computer business is a big, important, fast-paced industry whose aggregate annual income of $10 billion is in the same league with the telephone company's. The men in the industry are a breed apart. They are young enough, successful enough, and brash enough to think that all things are possible; one imagines that they are all 24 years old and working on their second hundred million.

In back of the computer people we have a group of entrepreneurs who would be delighted to set up the sort of communications network the computer people say they need, if only the government would let them go ahead and do it. For the most part their companies are not the names that made America great, but they make up for their lack of power and wealth with an unquenchable enthusiasm for the idea of one day running their own little backyard AT&T. They intend to achieve this in most cases by setting up a separate communications network based upon microwave links, satellites, community antenna television systems, or combinations of these.

In between sits the Federal Communications Commission. Long criticized for being understaffed, ill-equipped, and without the time, the technical competence, or the facilities to study the long-term consequences of important communications issues, FCC is a classic example of a regulatory body whose industry simply got out of hand. Amid the broadcasters, common carriers, mobile radio operators, computer manufacturers, Department of Defense, and outraged citizen groups, the FCC just has more than it can do.

For all of the fuss being made over it, there is little agreement about how much business the data communications market actually offers. AT&T says— and who should know better—that the present market accounts for 3% of their business, or roughly half a billion dollars. As for future growth, you can find estimates ranging all the way from 15% per year to 150% per year.

What everyone agrees on is that data communication is a big and rapidly growing business. Everyone agrees that the combination of a computer and a telephone line is one of the most fruitful collaborations since they joined a steam engine to an electric generator.

In fact, that is exactly the image computer people would like you to bear in mind. To appreciate the analogy, you have to begin by visualizing what society was like in the middle of the 19th century. Any industrial plant that wanted mechanized power had to set up its own steam engine. Then came electric power. Suddenly there was a simple and efficient way of transmitting power from a community power plant to individual consumers. Power utilities began to spring up and the luxury of power-driven machinery was available to everyone, not only big industrial plants, but small industrial plants; not only industry, but private homes.

Something of that sort is being envisioned for the computer. Let a computer talk on the telephone to some other machine and you have the makings of a new industrial revolution. No longer will the luxury of computers be reserved for the big industrial plants or even for the very rich. Now anyone who can afford a telephone can afford computer service. We are moving from a time when a computer is a large gray cabinet in an air-conditioned room, to a time when it is a plug in the wall alongside the electrical outlet.

They are beginning to call these things computer utilities. Although there is some quibbling about the name—in some circles "utility" raises the ugly specter of regulation—the concept is one in which a computer provides services to a large community of users conveniently and inexpensively.

Time sharing is one of the present forms of a computer utility. The latest directory of time-sharing systems shows that there are now about 150 systems in operation serving about 10,000 customers, with the industry not yet five years old. The business is expected to grow from $100 million to perhaps as much as $1 billion by the mid 1970s.

Time sharing is only part of the utility story. If communication charges are low enough, there is no reason why a company shouldn't use a remote computer for batch processing, information storage and retrieval, data distribution, and the whole range of processing activities that are now performed on company-owned machines. This kind of utility is particularly valid where the customer's activities are spread over a broad geographical area.

Add to all of this the traffic from remote polling, meter reading, shopping, alarm signalling, monitoring, banking, vending, voting, billing, ticketing, reserving, facsimile transmission—the whole range of functions that become possible with low-cost communication capacity—and you begin to get some feeling for the magnitude of the data communications market.

Just how much this will all add up to, and how concerned one ought to be about it, depends upon who you are and what crystal ball you are looking at. AT&T has been quick to clarify a statement made by chairman F. R. Kappel in 1960 to the effect that data communication on the network would "exceed in sheer volume the communication of speech" by 1975. What he meant, we are now told, is that the amount of information (presumably in bits) conveyed by machines would equal the voice information by 1975.

Perhaps as a result of the Kappel statement, perhaps for some other reason, it has been frequently stated in print and elsewhere that by 1980 half of AT&T's revenues will come from data. That is greatly exaggerated, says AT&T. By 1980, the figure will probably be no more than 5% to 10% according to William Quirk, AT&T's data communications marketing director. However, in an industry-wide study by Stanford Research Institute, the conclusion was that "by 1980, between 10% and 50% of the Bell System may well be serving data users."

The predictions you make obviously determine whether you are going to be concerned over the possible scarcity of communications capacity. To many, AT&T seems to be taking the position that a service accounting for 3% of the business deserves about 3% concern. Moreover, they state with great confidence that they will have the plant necessary to handle the load. Lee L. Davenport, president of GT&E, the second largest telephone company, was quoted in *Electronics* as saying, "The computer people foresee tremendous growth in data traffic. However, I think their forecast is overly optimistic and I believe that we are successfully keeping up with the rate of growth."

Computer people are particularly concerned by the difficulty the telephone company is apparently having keeping up with the voice service in New York and a few other large cities. "If there are already places where they can't handle the voice traffic," they say, "how will they ever handle the data traffic?" There has also been some speculation that increasing data traffic is, itself, partly responsible for the present service difficulties. AT&T feels that this is probably not a very significant factor. But neither the telephone company nor the computer people will soon forget the experience at the recent Fall Joint Computer Conference in Las Vegas where computer conversation from machines on the exhibit floor and voice conversation from an unusually large number of talkative visitors so badly tied up the Las Vegas telephone plant that certain areas of the city couldn't get even so much as a dial tone for days.

There is little question, however, that the telephone network could be made to handle any load of voice and data traffic ever likely to be put on it, provided the telephone company prepares itself for the traffic in advance. In terms of capacity, the telephone system is almost infinitely expandable; it was designed that way. And what with high-capacity coaxial cables, circular wave guides, communications satellites, lasers, electronic switching equipment, and a number of other recent advances in communications technology, the opportunities for expansion have never been greater.

It is not capacity that troubles computer people—nor even so much AT&T's preparations to increase capacity. What troubles computer people is that the telephone network is not, and may never become, anywhere near the most efficient or inexpensive medium for the communication of data available with modern technology.

And for a closer look at that question and its ramifications, we will have to travel over to the Federal Communications Commission.

You get off the elevator on the second floor and follow the corridor around to the right until you get to the Public Reference Room. Although the building is relatively new—one of those bronze-colored affairs with exposed columns that threatens to become the new standard of Federal architecture—the FCC's Public Reference Room looks as if it had been lifted right out of the old FCC building, wherever that was, and installed—walls, tables, and furniture—right into the new building.

You go to the desk and ask for Docket No. 16979. The girl disappears and returns carrying four thick, green-bound volumes with the number 16979 stenciled plainly on the front. She will, if you insist upon it, deliver the entire docket which, when it arrives, will stretch some 6 ft. along the counter. Upon examination, you will find that each of the 10 volumes is stuffed full of documents ranging anywhere from one page to 300 pages in length. Welcome to The Maze.

What you have before you is the response to the FCC's inquiry into Regulatory and Policy Problems Presented by the Interdependence of Computers and Communications Services and Facilities (and Supplement), November 10, 1966, and March 2, 1967. It is second only to a house of mirrors for fascination, bewilderment, and frustration.

It is all there, argument, counter-argument, additions, appendices, the whole thing. Despite the fact that it was mostly written by lawyers—or in any case submitted through their offices—it is quite readable, free of jargon, and not above the occasional amusing touch. If you read closely, for example, you will see where Ma Bell, irrepressible schoolmarm, raps a competitor lightly on the knuckles for the grammar in his argument, and another place where one of the data processing trade associations takes a gratuitous swipe at Western Union for their unadventurous management.

The frustrating part is that there is no way for an outsider to draw a clear line of justice between the positions represented. For every ironclad argument on one side there is an equally ironclad argument on the other. Read two documents side by side and you cannot help but feel that there is a good deal of common sense and logic in both.

Even the FCC blanched when it came to the monumental job of reading and deciding upon all the questions they had asked. So they hired Stanford Research Institute to read it for them.

Six months and a quarter of a million dollars later SRI delivered up its summary and analysis in seven volumes which, being printed and bound rather than mimeographed, were skinny enough to be slipped into the back of the 10th volume of the docket. Still not satisfied, the FCC asked for comments on the SRI summary and got another load of paper work. This time it wasn't so bad: only one volume and a little over. And now with Docket No. 16979 inching up toward 12 volumes, we have a nice tidy stack of paper . . . but still no decision.

It is no wonder, really. The questions the FCC is being asked to wrestle with are enormously complex. This, for example:

In the old days, when a telegraph line was a wire with a key at one end and a sounder at the other, the process the telegraph company used to route a message from a particular sender to a particular receiver was known as "message switching." That is, when the telegraph office in Cleveland sent a message to the telegraph office in Chicago, the message carried instructions for delivery to its eventual destination in Rockford. What the Chicago office did, in that instance, was called message switching—it routed the message from one place to another according to the instructions given by the sender.

Today, the telegraph company still routes its messages by the same process, except that the actual switching is now done inside a computer and is therefore difficult to separate from other things that go on inside a computer.

Although the process has changed some, the telegraph company remains understandably adamant about one thing: For the protection of the customer and the preservation of network integrity, the process of message switching must remain the exclusive privilege of the licensed common carrier. As with the U.S. mail, no one but an authorized carrier is allowed to handle the message until it gets to the intended receiver.

No one is arguing with this principle. What people argue about is precisely what activities should be classified as message switching.

Suppose, for example, that I call you and tell your secretary that I will meet you for lunch at Luigi's. Is your secretary doing the message switching? Well, no, as far as the carrier is concerned, the message was delivered when it got to your secretary. But suppose you have an answering service? Does the answering service do message switching? Or suppose you have a tape recorder attached to the telephone lines. Is that message switching? As it happens, in both of those cases, it has already been ruled that neither of these constitutes message switching and therefore neither of these needs to be reserved as a regulated activity of a common carrier.

But now, suppose I am Standard General Universal Corporation in Chicago and I am sending a big collection of operating data to our computer in New York for processing. I am also asking the computer to relay a copy of the data on to seven other companies there on the East Coast which also happen to be hooked to our computer.

That is when the fuse blows at Western Union.

They maintain that you are doing message switching, and message switching is the exclusive function of the regulated common carrier. It's not just a matter of sticking to the old rules. If they allow you to distribute a message from your computer center, there is nothing to stop you from doing this on a large scale and eventually setting up your own private communications network in competition with the telephone company. Or more to the point, if they allow you to do the distribution, they will be denied revenues that are rightfully theirs, either for the distribution of that message from their message switching center in New York, or for the distribution of seven separate messages from the office in Chicago to the seven companies on the East Coast.

There are infinite variations and subtleties that make such questions difficult. Should there be a distinction between message switching on leased lines, as in the example above, and message switching on dial-up public lines? Or if the message being switched is processed in the computer—changed, somehow —before it is distributed, is that still message switching? Is there a difference between the sort of message switching I have described above and the sort of thing that would occur if I merely stored the data in my New York computer so that it was accessible to each of the seven companies?

Only a communications lawyer can thread the labyrinthine paths by which such a seemingly harmless and innocent activity gets escalated through a hierarchy of logic to the point where it threatens the foundations of the business.

So complicated is it that even AT&T has slipped a time or two, or so it is claimed. And even that depends on who is keeping score.

Nearly all of the issues being explored in the FCC's inquiry get into intricate legal complexities of the sort encountered in message switching. Indeed, most of them are even more complicated.

But complexities aside, the basic points of controversy between the telephone people and the computer people are reasonably simple to describe and understand. The SRI summary listed ten of them. Most of these, however, fall into two categories: There is the Digital Question, and there is the Variable-Channel-Size Question.

The digital question is potentially the most damaging to the telephone company. To begin with, even a telephone engineer would agree that a computer and a telephone line are not natural allies. Telephone lines were designed to carry the human voice while computers talk in pulses of electrical current. Or to say it another way, telephone line is inherently an analog creature that carries an electrical analog of the human voice. A computer is a digital device that deals in bits coded in some way so as to represent information. The two don't get along well together. Send a string of bits for any distance down an ordinary telephone line and it comes out looking as if the mice had been at it.

To make them compatible, the telephone company provides a box they call a data access device, or what is known generically as a "modem" (for modulator/demodulator). There are all sorts of modems, but the general scheme is to "modulate" the digital data into the sort of language a telephone line understands. At the other end, it "demodulates" the telephone language back into pulses. The tones you hear when you use a push-button telephone are an example of the kind of modulated pulses a modem might send out.

What computer people would like, given the opportunity, is a digital network. This, they claim, would be cheaper, partly because it avoids the expensive modem, some of which cost upward of $10,000 each, but also because the error rates could be reduced from, say, one error in 10^4 bits to one error in 10^7 bits. In addition, the digital signal would require less communications capacity on a digital network than it does on an analog network.

That comparison can be deceptive however, for the key question is not how much communications capacity or bandwidth the signal requires, but how much the carrier charges to carry it. That may, or may not, be related to bandwidth. After all, the telephone company is already finding it profitable to convert voice signals into digital signals through a process known as pulse-code-modulation but at a cost of using up something like 15 times the bandwidth of the undigitalized voice signal. The economy it turns out is in the multiplexing and switching mechanism at the central office, a saving that more than makes up for the prodigious waste of transmission space.

In addition to economy, the attraction of a digital system—whether it is carrying machine language or coded voice—is that the information can be transmitted without being degraded by noise en route. This is because a pulse is a pulse regardless of how badly it is deformed by noise or other transmission irregularities; as long as it is recognizable, it can always be regenerated as a nice clean pulse again, indistinguishable from the pulse that left the transmitter.

Although the Bell System now has some 8 million miles of digital voice grade circuits out of a total of some 700 million miles, none of these digital circuits presently has a run of over 50 miles and none of them is used in strictly digital mode by the customer. Ironically enough, digital data destined to be carried on the digital circuits must first be converted to analog signals and then reconverted to another kind of digital signal before they are sent on the cable. AT&T has offered to provide end-to-end digital service, but so far they have had no takers.

The variable-channel-size question is somewhat simpler although it may not be any easier for the telephone company to accommodate. What the computer people would like is a network that provides them with only as much or as little communication capacity as they really need for any particular application. The telephone network, being designed for voice, can do no more than offer a voice channel, which has a capacity of 2,000 bit/second with the standard modem. If you need more than this, there are some special services available, particularly on leased-line basis, but if you need less than this, or if you need more in one direction than in the other, or if you need an amount that falls in between AT&T's widely spaced offerings, you have no choice but to accept the next largest capacity and let all the unused capacity go to waste.

Put these two questions together and what you have is a question of economics. Data processors contend that their communications costs could be cut enormously using a system designed for—or at least carefully adapted to—the needs of data communication. AT&T argues that this strikes at the basic assumptions upon which the communications network is built. The advantage of a large general-purpose communications network, they claim, is that it can offer a variety of services to a variety of customers at a price that is reasonable for all. To provide a special service to a limited group of customers at special rates only increases the rates the others have to pay. And letting some other carrier provide that special service is described by them as "cream skimming."

In the last several years, partly as a result of the FCC inquiry, and partly as a result of other things, AT&T has shown a surprising willingness to relax some of its long-held and dearly cherished restrictions on the use of their network.

For example, AT&T has offered, on a trial basis, a 1-minute minimum hold time in place of the 3-minute minimum on calls made between 11 P.M. and 8 A.M. anywhere in the country. AT&T has always maintained that the cost of connecting and billing their customers is such a large part of the cost of communication that they needed a 3-minute minimum to assure recovery of their costs. The trial, instituted largely at the instigation of computer users who want to send short bursts of data, has attempted to determine whether the increase in telephone usage will offset the loss in revenue per call.

AT&T is also offering a great many more channel capacities than were available in the past. In addition to the normal 2,000 bit/sec switched-network voice channels, the Bell System now also offers wideband dial-up service with a capacity of 50,000 bit/sec within and between four major cities. Moreover, when videophone service is made available on a nationwide basis in a few years, it will provide a switched network with a channel capacity of about half a million bits to begin with, and perhaps something more than 1 million bit/sec as the modems are improved.

A great many options are now also possible with dedicated private line service; there, the Bell System will provide channels varying in capacity from the capacity of a voice channel up to 250,000 bit/sec.

The Bell System has also become more lenient about sharing lines. In the past, it was prohibited for a Bell System customer to share his lines with someone else. The reason is obvious. If I have telephone service, I could offer to string a wire over to your house and install an extension phone over there. That, says AT&T firmly, is reselling regulated service and is strictly disallowed.

The Bell System now offers a variety of sharing arrangements on private lines, and has set up an arrangement whereby several people can share services on the switched network. It is, of course, set up in such a way that there is no danger of a customer's putting himself into the telephone business. These sharing arrangements are essentially the same as offering the customer a variety of channel capacities to choose from.

Perhaps the most far-reaching change in recent years is the celebrated Carterfone decision of June 1968 in which the FCC ruled that AT&T's long-standing prohibition against connecting anything but telephone company equipment to the telephone line was "unreasonable and unduly discriminatory." It is one of the landmark cases of communications law.

Thomas Carter, a Texas inventor and manufacturer, produced a device in 1965 that connected two-way mobile radio stations to the telephone network. The telephone companies claimed it was illegal. After three years of argument and counter-argument, Carter finally won.

Suddenly, one of the most jealously protected principles of the telephone company operation had been breached; suddenly, anyone who wanted to do so could attach his whatsis to the telephone lines and use the $45-billion facility in any way that suited his convenience.

With certain restrictions, of course. The customer is still not allowed to use his own device to control network switching. Moreover, anyone who wants to use the telephone lines in some unusual way must do so through a device that attaches to the end of the telephone lines to protect the network from spurious signals, noise, overvoltages and the like.

Although the telephone company fought the decision, they did not contest it once it had been made, and have since come around to the view that it might actually be a good thing. "We will benefit from this because of the increased use of our lines," says AT&T's marketing vice president, William Ellinghaus.

The result of all these changes is that the carriers have now given the computer people at least some of what they asked for in their responses to the FCC inquiry. And they will continually try to adapt to changing requirements. "We are carrying on a continuing dialogue with the data processing people," explains Ellinghaus. "We are aware of their needs. In fact, I don't know of any customer request for provision of data facilities that we haven't met."

Still, the telephone network is never likely to become all-digital for the benefit of data processors, nor is there likely to be any significant retreat from the position that costs should be averaged between inexpensive facilities and costly facilities, or that tariffs should be averaged between profitable runs and unprofitable runs.

In some ways, change is easier for the Bell System than it is for other big systems. Large as it is, it is all owned and run by essentially one organization. It controls the services, the equipment, the manufacturing operation, and to some extent the customers. It is not like the automobile industry where a small change by one major manufacturer has repercussive effects on numerous other big and little businesses, most of which are beyond the control of the automaker, as Donald Frey pointed out in his earlier chapter.

Then too, the telephone system was designed with a certain amount of versatility built into it; a change as small as the switch to a 1-minute minimum holding time, or a line-sharing arrangement, can be accomplished with some rather inexpensive changes in the switching-office machinery.

But change does not have quite the same meaning to a telephone man that it has to a computer man. One of the important factors in the present confrontation between telephone people and computer people is that they represent two quite different schools of engineering. The computer people have seen their industry torn up by the roots three times within the last two decades. The equipment they design and build has a technological lifetime of a half-dozen years ·at most, never mind its actual operating lifetime. Computer engineers are accustomed to designing systems with equipment that becomes rapidly obsolete.

Telephone engineers, on the other hand, are the acknowledged masters of the long-lived, ultra-reliable piece of machinery. So what if it is overdesigned by ordinary standards. So what if the design is conservative. Nothing else would work in that environment. Imagine designing a telephone handset that couldn't be dropped on the floor at least once a week! Imagine designing a dial that couldn't survive a dash of soapy water now and then! Imagine building a switching relay that malfunctioned once in a million times! The service and maintenance people would never go home.

But more than that, telephone engineering is a very exacting, close-to-the-margin kind of business. Save a penny on this device and multiply it by so many devices, or by so many hours of operation, and you have a sizable number.

Take switching office equipment, for example. The chief engineer of a switching office is responsible for providing precisely the right amount of switching gear to handle the load, no more, no less. The rule is (or was when this was explained to me some years back) that only 1 out of 100 calls during the busiest hour of the day is to encounter a 3-second dial tone delay. It is a very tight specification. If the delay is 4 seconds, the engineer is criticized for giving his customers poor service; if the delay is never more than 2 seconds, he is criticized for wasting the stockholder's money by providing more machinery than is necessary.

This bears on the computer question because a computer doesn't have the same telephone habits as a telephone subscriber. According to telephone company statistics, the average telephone has a use-factor of about 3 minutes per hour. All the switching machinery calculations are based on this assumption; it's all very carefully worked out. Now comes a computer that gabs on the phone for hours at a time, perhaps even days. A switching plant that has many computers for customers is bound to give a telephone engineer fits.

Added to all of this is the requirement for compatibility, surely the most inhibiting and potentially stagnating restriction to be forced upon an engineer. It can be challenging, yes, this business of making all the stuff you design today work compatibly with all the stuff you designed 40 years ago. But if you already have close to a billion dollars invested just in telephone instruments, if you already have thousands of miles of coaxial cable, hundreds of microwave towers, and 700 million miles of circuits, all devoted to doing the job one particular way, the challenge is eventually met only through great inventiveness, if at all.

And the challenge gets more difficult every year. Every year you are spending more and more engineering effort trying to make the new attachments fit the old Electrolux. In that environment, the old way of doing things hangs on for a long time. Is it any wonder that Western Electric is still making new step-by-step switches, and that the Bell System is still installing new step-by-step switching offices, despite the fact that the design dates from the early part of the century and has presumably been obsoleted not once, but three times: first by the panel switch, then by the No. 5 crossbar, and finally by the electronic switch?

Is it any wonder that computer engineers and telephone engineers can exchange ideas without ever once realizing they are not talking about the same thing?

One of the most powerful restraints on change in the telephone company at the present time is not engineering, nor tradition, nor conservatism. It is capital. Because of the natural growth of the voice network, AT&T must raise between $2 and $5 billion of new capital each year. The biggest share of this money cannot be spent to provide new services, but must go toward providing the old services to a rapidly growing group of new customers. Raising $2 billion is difficult, particularly when the company's rate of return is strictly limited by government regulation. If one needs to raise a half-billion dollars for a new communications service, it may be a lot easier for some independent than it is for the telephone company.

Although Western Union faces some of the same difficulties, they have chosen to approach the data communications market from a somewhat different direction.

One might have expected that Western Union, being the nation's only domestic "hard copy" or "record" communications service and, therefore, holding a virtual monopoly on nonvoice communication, would have been in a position to pick up all the data communications marbles. But they did not and are now striving desperately to make up for lost time.

Beginning in 1966, Western Union began to augment its traditional telegraph and teletypewriter (Telex) service with an array of new communications services. The company grandly calls the new services "computerized communication" and, in the best computer traditions, invests them with names such as SICOM, INFO-COM and TCCS. Essentially, what computerized communication means to Western Union is that message switching is done by computer. It also means that messages can be stored in the switching computer either for purposes of keeping records, or for delivery at some later time.

Brokerage firms are using these new services to keep records of all the transactions that take place between branch offices and the main office on Wall Street. The storage capability also makes it possible to queue up messages when the teletypewriter at the destination is busy, or send a list of addresses along with a message to the switching center computer and have the computer distribute them automatically.

Western Union wants to add remote data processing, either at their switching center or at a separate facility. In this way, a customer could buy all of his computer and communications services—remote batch processing, time sharing, message switching, communication, the whole package—all from one source.

That is the basis of a very heated dispute.

Although regulated common carriers are not presently prohibited from offering unregulated data processing service (General Telephone and United Utilities both operate data processing centers) it is one of the questions that is being very closely examined by the FCC. It was, according to the Stanford Research Institute summary, "easily the most controversial issue" in the FCC's inquiry into computers and communication.

The argument against it is that the firm might use the profits from its regulated service, to subsidize its nonregulated service, thus preventing or inhibiting competition. Mixing of regulated and unregulated activities in one company is not unknown—ITT, RCA, and several others, conduct both regulated and unregulated activities—but it is thought that, in this case, because of the strong interrelation between computers and communication, it would be difficult to keep the two activities separate. The SRI summary concluded that because "Accounting categories are inherently arbitrary and manipulable . . . a clean separation of costs and profits [for the two functions] is simply impossible." In a summary of the summary authored by one of the SRI investigators, it was claimed that "Absolute prohibition [of carriers into data processing] would be easy to administer and would meet with enthusiastic reception by the data processing industry."

SRI was not enthusiastically applauded, however, for adding that Western Union represented a special case and should not necessarily be forbidden to do data processing. Western Union's situation, they said, "is such that the danger of predatory conduct is negligible."

Yet without some form of governmental benevolence, Western Union's growth could be severely limited. If it should happen, for example, that unregulated data processors were allowed to do message switching, and if, in addition, common carriers were forbidden to do unregulated data processing —both of which are reasonable possibilities—Western Union might be left with little to offer beyond a boy on a bicycle delivering messages glued to pieces of yellow paper. No one, not even its competitors, wants to condemn Western Union to that.

Regardless of what happens to Western Union, data processors will still be left with many of the same complaints. It is for this reason that they have begun to look elsewhere for new communications services.

In any other circumstances, Jack Goeken would be called an eccentric. He has a dream of achieving "the Impossible," and he has a lurid collection of stories about how the "pressure of large organizations" has prevented him from doing it. He has (or had) little money, little formal schooling, and little real understanding of how to attack his particular problem. From all appearances, there is little to separate him from the designer of the perpetual motion machine, the discoverer of antigravity, or the inventor of the 150-mile-per-gallon carburetor.

Except that on August 14, 1969, the FCC made Jack Goeken a very respectable eccentric. It was on that day that his company, Microwave Communication, Inc., was awarded a licence to set up a common-carrier microwave communications system between Chicago and St. Louis in competition with AT&T. It was on that day that the impossible dream came true.

Ten years ago, John D. Goeken was a sales and service representative for General Electric in Joliet, Ill. "We had begun with a radio and TV service shop. Later we got an FAA license to service aircraft radios. Then we picked up the GE franchise and branched out into two-way radios selling them mostly to truckers."

In 1963, Goeken and four friends got the idea that truckers would be able to make better use of their radios if they could communicate with their home offices while en route between Chicago and St. Louis. He and his friends decided to set up a microwave link between the two cities that would be accessible to two-way radio at any of the microwave towers along the route. The five of them chipped in $600 apiece with the idea that it would probably cost about $3,000 in legal fees to get the license.

Six years later, only two of Goeken's four partners are still in the business and they are not active participants. He has now used up the original $3,000 and another $397,000 in addition. Jack Goeken is a very persistent man. "I don't think I would have done it," he says, "except that they were so anxious to prevent me from doing it."

"They" is the common carrier, principally AT&T. He loves to tell the story about his confrontation with the telephone people before the FCC. And well he should; it is one of the rare instances of a battle between Private Individual and Big Corporation in which Private Individual won.

"When I arrived, the telephone company already had their lawyers lined up. There were thirteen of them. They come down the bench around the corner and onto the side to where I was sitting. There I was, just me and my lawyer. Behind the line of lawyers were more telephone representatives—technical consultants, engineers, scientists, communication experts."

The main thrust of the AT&T attack, according to Goeken, was an attempt to discredit his application on technical grounds—to picture him as a small-time operator with little understanding of the technical problems who simply wanted to slap together a quick and dirty communications link. Goeken, on the other hand, claimed that it was unfair to compare his system, which he believed to be thoroughly adequate for the job, with the more-than-adequate, overengineered—what he calls "gold plated"—AT&T designs.

He describes the difference this way: "When AT&T comes in to set up a microwave tower, they buy a quarter section of land and put a big chain link fence around the outside. Then they put up their tower and build a heated, air-conditioned, blast-resistant building at the base to house all the equipment. By the time they get all done with access roads and everything, for all the sites they would need between Chicago and St. Louis, they would have a million dollars or more invested.

"I have a different idea. Most of our tower sites are in the middle of farmers' land and I've talked to a lot of these farmers. They understand my problems and I understand theirs. My idea is to lease from them a small piece of land at the intersection of two fences. We put up our tower at the intersection and run our four guy-wires out along the fence lines. This way I can lease all my tower sites for $170 a month and the farmer doesn't lose any farmable land. They've even offered to help me dig the trenches for my power cables. Why should I go out and buy a tractor when I can borrow one from the farmer?"

AT&T—giant of industry, leader of technology, ruler of communication —was understandably aghast. "Whoever heard of a microwave tower with four guy-wires," sniffed one AT&T witness, fumbling for a defense. "Microwave towers are always designed to have three guy-wires. And if he uses three guys, he has to put two of them out in the middle of the farmer's field." Another criticized the small transmitter building Goeken planned to build at the base of his tower. "It's not hurricane-proof," said the witness.

"How many hurricanes do you get in the middle of Illinois," replied the examiner.

More serious objections came from the witness who questioned Goeken's placement of towers. "The plans call for some towers to be placed as much as 28 miles apart," said AT&T's technical expert. "Anyone who understands anything about the effects of rain attenuation and other atmospheric disturbances would never put microwave towers more than 22 miles apart, 24 miles at the most."

This was a serious criticism. "What chance has an Illinois farmer like me got against expert testimony like that?" Goeken explains in telling the story. "Who should know more about the placement of microwave towers than AT&T?"

Although he may act a little like an Illinois farmer, and may even represent himself as one at times (he is actually an Illinois preacher's son), he's not so easily put down as all that. Next day he arrived at the hearing with information showing that some of AT&T's own towers in Illinois were as much as 36½ miles apart.

"Where did you get this information?" said the examiner.

"Upstairs in the office of the Common Carrier Bureau," replied Goeken.

So in the end, Jack Goeken got his license. An appeal by AT&T slowed progress for a time and accounted for a certain amount of caution by his investors, but the appeal was denied and Goeken has now begun to build.

Goeken originally planned for his Microwave Communication, Inc. system to carry 300 voice channels. The first week after the FCC decision he had already sold 50 channels just on the strength of publicity, and he has since sold about 250 more. He has therefore increased the capacity of the initial system to 600 channels, and included provision for increasing the capacity to 1,800 channels.

The service being offered by MCI, is a dedicated private-line point-to-point microwave service which will carry voice as well as data, between the two cities. Although it is designed to interconnect with the Bell System at each end to provide distribution to its customers, it will not allow access to third parties. That is, there will be no facilities for an MCI subscriber in Chicago to call some friend in St. Louis. It is not an alternative to the dial-up service and does not compete with it. It is a competitor of the common carrier's private-line service.

As such, it offers a number of features not offered by the carriers. The subscriber can buy any of 138 possible information-carrying capacities, from 200 Hz up to 960,000 Hz. He can get one capacity in one direction and another capacity in the other direction, if he wants it. He can also get reduced rates for half-time service. The result is that he can buy communication service between St. Louis and Chicago at rates that are significantly lower (between 55% and 92% claims MCI) than the service offered by the Bell System.

Nor does MCI intend to stop with the Chicago-St. Louis route. They have already established an affiliate company, called Microwave Communication of America, or Micom. Micom will undertake the job of setting up a nationwide network of microwave communications links. Applications have already been filed with the FCC for routes between Chicago and New York, New York and Boston, New York and Washington, San Diego and Seattle, and a number of other places. Ultimately, there will be filings for routes that will provide a complete nationwide network.

It is not known that the FCC will approve the new applications. The vote on the original MCI decision was four to three, and the wording of the majority opinion was such as to indicate that it was experimenting with competition as an alternative to regulation. Many observers feel it is not likely that FCC will try another experiment until they have had a chance to see how the first one turns out.

That attitude, if it exists, may slow down the decision on other applicatiins now pending in the FCC. The most ambitious of these is the application by University Computing through their subsidiary, Data Transmission Company, for a nationwide data communications network.

The Datran network differs from MCI in several important respects. It is, first of all, an all-digital network, whereas MCI is strictly analog (at least for the present). The technical difference is that the repeaters stationed along the transmission path are not simply amplifiers, but are pulse regenerators. The operative difference, as explained earlier, is that it permits efficient transmission of pulses, low error rates, and reduces noise almost to the point of extinction—all very attractive features for the data communications customer.

It does mean, however, that Datran cannot interconnect with the Bell System without sacrifice of the all-digital feature. Datran must arrange for its own connection to the customer's premises. It plans to do this by short microwave links, infrared transmitters, or cables, whichever is most appropriate.

The second important difference is that it is a switched network: Anyone on the Datran lines can exchange data with anyone else on the Datran lines.

Like MCI, Datran will offer a wide range of information-carrying capacities, but unlike MCI, the information will be strictly limited to data. Because they are not dependent on the Bell System, they will also be able to offer fast connect and disconnect times, a very attractive feature for data processors who want to send only short bursts of information.

Neither Datran nor MCI claim to be in competition with the other; both seem confident that there is enough variety in the communication market to support everyone . . . including the telephone company.

The most impressive evidence of the seriousness of Datran's bid is the elaborate mathematical model they have fashioned to help them estimate the size of the data communications market. Not satisfied with those all-encompasing projections that have data communications growing by so much per year, or so much per decade, Datran has broken the whole thing down into separate questions: Who are the data communicators? What kinds of data? How many? How fast? Between what cities? etc.

With their model, they can forecast the need for data communication in any one of the seven major data communications industries—banking, manufacturing, petrochemicals, insurance, transportation, securities, and retail. They can pull out enormous books of computer printout and find that in 1974, for example, communication by banks between Cleveland and Nashville at 4:00 of a typical morning will require such-and-such capacity in the speed range of so many bits per sec. They can then calculate maximum trunk size, optimum switching plant locations, and the Datran share of the resulting revenue. All very neat. And if the real world works anything like the world they have tucked away in their computers, they should all be millionaires the month after FCC gives them a license.

Ay, there's the rub. The carriers have protested Datran's filing, chiefly on the grounds of the "cream skimming" argument. They claim that this competition draws away their profitable business and leaves them with the less profitable business. Consequently, there is less money for cross-subsidization;

tariffs for ordinary telephone service must increase. A common carrier competing with the existing common carrier, AT&T concludes, is not in the public interest.

All of this has been heard before, of course. What's more, it has always been a successful argument until the MCI decision. It is that breakthrough that gives Datran hope.

Both MCI and Datran build their case for a separate network on the grounds that they are offering services not presently offered by the telephone company. They are not cream skimming, they claim; they are not even competing; they are merely opening up new markets.

E. A. Berg, Datran's vice president for operations, further argues that if there is any cross-subsidization going on between data communicators and voice communicators, the money is flowing the other way. According to Berg, much of the data communication—time sharing, for example—uses only local circuits. "The data communicator pays just as much for his local call as you and I do . . . and he may tie up the lines for hours at a time. The man who's paying the subsidy is the man who uses his telephone for five minutes a day and pays the telephone company the same as a machine that uses it five hours a day. That's where the cross-subsidization is."

Datran has made things particularly tough for their application, though, say some observers, by being a wholly owned subsidiary of University Computing, a large time sharing and data processing service organization. In asking FCC to rule on the application for a national data network in competition with the present carriers the question inevitably arises of whether a regulated common carrier should be allowed to do unregulated data processing.

Lurking in the wings all this time is CATV, dark-horse candidate for data transmission. Community Antenna Television or cable TV is that system of high-gain antennas and coaxial distribution cables that brings TV programs to communities too far out in the sticks to pick up reliable signals with ordinary rooftop antennas. Or at least that's what it started out to be. What it is slowly becoming is a wideband communications network capable of carrying into every home in the country thousands of times the information carried by the present telephone network.

Again, there are complex economic and regulatory forces that have shaped the first stages of this transformation, but the most important ones are a general friendliness toward CATV on the part of FCC and Department of Justice, and a realization on the part of viewers that CATV gives a better TV picture and a more diverse variety of programs. Together these forces have caused the 20-year-old CATV industry to grow from half a million subscribers and 600 systems in 1960, to nearly four million subscribers and 2,500 systems by the end of the decade. Moreover, many of these are now in large cities—San Diego, New York, Denver—where the combination of clearer signal and additional channels is considered well worth the $5-per-month subscription fee.

It needs to be emphasized, however, that there is a very significant difference between CATV and the telephone network. Telephone service is established through a switched, person-to-person network whereas a CATV system is merely a distribution facility—everything that goes in the transmitting end is available to everyone connected to the receiving end. It is more like a giant party line than a network of private communication paths.

There was a time when that was a very important difference. It is one thing to lay a pair of wires from a central office to a community of, say, 500 subscribers and connect them all to that same pair. It is quite another thing to lay a separate pair of wires to every house in the community and bring all 500 pairs back to the central office. Added to that is the problem of switching between them in such a way that a man on one pair can reach a chosen man, and only that man, attached to another pair.

But, say some, there may be other ways of doing this. Suppose, for example, you and I are both connected to the same CATV cable. With the help of the appropriate logic and coding circuits, there is no reason why I can't exchange information with you, and you exclusively. I simply arrange for all of the information I send you to carry a "header"—a series of pulses, say—coded in such a way that it automatically opens up your receiver. The information you send back to me is preceded by a header coded to unlock my particular receiver.

If the coding and detecting machinery is made to function automatically once our connection is made, we should be able to use this distribution network exactly as if it were a private line. In fact, it has been suggested that some years from now it may be more practical to design a switched person-to-person network in this fashion rather than in the old separate-wire telephone-network fashion.

That, however, is still some distance in the future. The present attraction of the CATV network is the information carrying capacity that is available. The cables now being laid have a theoretical bandwith of close to 300 megahertz, enough capacity to carry 100 television channels or 100,000 voice channels. Because the repeaters are not quite up to the cable, however, the actual capacity of the best systems is more like 20 TV channels.

Some CATV systems are now being installed with a capacity of 40 TV channels along with the capability for two-way communication so that the subscriber can send information back to the transmitter. There is enough capacity in just one two-way video channel (actually two channels on the cable) to send an enormous amount of data, for it is roughly equivalent to

1,000 telephone lines. This could easily be enough to serve a large number of industrial subscribers if the appropriate channel sharing machinery were to be made available.

It is this prospect that makes CATV networks potential data carriers in the near term. A hesitant step in that direction was taken when Communications Properties, Inc., operator of 13 CATV systems in Texas, filed an application with FCC to use their interconnecting microwave links for the communication of data. In the filing, there is no mention of interconnection between the microwave links and the CATV systems, but CPI is said to consider that a definite future possibility.

In the long term, CATV is viewed as a potential medium for distribution of computing power into the home. Computing power, in this case, means more than just a time shared computer terminal; it means information retrieval, electronic mail delivery, individualized entertainment and education, and a variety of services frequently lumped under the heading of "demand broadcasting." It is quite likely that some of these services will be carried on the switched, wideband videophone cable rather than on the CATV cable, but no one is yet making any predictions about just where the line between those services will be drawn ten years from today.

Domestic communications satellites add another layer of complexity to an already complex matter, but their impact on data communication is not expected to be so great as upon other forms of communication, particularly television networking. The reason for this is that a satellite is most effective when it is used to connect one point with many other points. To do that kind of distribution with land lines would be very much more expensive than it is with satellites. And to employ satellites in applications where this inherent feature cannot be exploited is to waste some of their potential.

Or so it has been said. But even this is a matter of some dispute. The plain fact is that there has not been enough experience with satellites and their economics, particularly on a domestic level, to conclude anything much about their potential usefulness.

Early in 1970, President Nixon's communications advisory group recommended to the FCC that the bidding for a domestic communications satellite be opened to private interests on a competitive basis. Since 1966, Comsat has had a proposal for a domestic communications satellite before the FCC, but the proposal has never been approved or denied. The recommendation would not exclude Comsat, but would simply put it on a competitive level with everyone else. Perhaps an answer to the much-discussed domestic satellite question may not be too far away.

A director of planning for one of the large computer manufacturers was recently quoted as saying that he foresaw the possibility of this all coming down to a three-layer network; one for business data, one for consumer data growing out of CATV, and one for telephone service. That seems possible. Electronic Industries Association has filed a paper with FCC favoring a two-layer network: a switched videophone service and a wideband CATV-type information service. That seems possible too.

In fact, the most striking aspect of the new vitality in communication is to make a lot of things seem possible that seemed impossible not many years ago.

In the end, the burden of choice must fall on the FCC. Between the idea and the reality, between the conception and the creation, falls the shadow.

It is a position of responsibility many believe to be beyond FCC's present capabilities. It has even been suggested, perhaps more in desperation than anything else, that the best way to solve the problem will be for the government to remove all restrictions and regulations. The resulting chaos, it is thought, would quickly solve problems that might take decades to solve otherwise.

A more likely course of action is beginning to take shape in Washington with the proposal for an Office of Telecommunications Policy in the executive branch. It would replace the Office of Telecommunications Management established some years ago, which has been criticized as being a generally ineffective organization. The new Office would presumably take over responsibility for certain FCC functions, but more important, would provide the detailed research into questions of policy presently lacking in the FCC.

Meanwhile, there is faint uneasiness to be detected around the executive offices at AT&T. For the first time in the company's history, disquieting noises about the quality of service are beginning to filter up from the street. For the first time, critics from the technical community are beginning to talk bitterly about "an unwillingness to respond to the forces of change."

It does not go unnoticed, this talk. The phrase, "acting only to protect her own monopoly" cuts very deep in an atmosphere where even the word "monopoly" is considered impolite. The model public utility that has worked so hard to achieve a standard performance against which all other utilities are gauged shows herself to be unusually vulnerable to even a suggestion of the idea that a big, benevolent, clean-living supporter of the public interest may not really have the public interest at heart after all.

"I hope people still think of us as the Good Guys," one of the AT&T executives told me bravely.

Frankly, I do too. But as it applies to this matter of data communication, I can't help remembering what Durocher said about Good Guys.

March 1970

5

How a Young Industry Changes

Computer capacity once looked like a golden egg that could be sliced and sold by anybody knowing the technology. In less than 4 years entrepreneurs, most of them technical men, have expanded the time-sharing industry to more than 140 companies. Now many of these companies are discovering—though still not understanding—the troubles in their technical and business strategies. Evan Herbert explains what led up to the current shakeout, then takes a look at one company whose strategies seem to be paying off.

Mr. Herbert is a Senior Editor of *Innovation* Magazine.

Today there are about 140 companies in the U.S. that simply could not have existed at all before the mid 1960s. All of them have in common new way of using computers, popularly called time sharing. All of them now face in common a shakeout of their brand-new industry. This comes relatively earlier than the similar classic shakeouts in other technology-based industries such as automobiles and broadcasting.

But in a way, the computer time-sharing industry can't really be compared with either of these. It does not manufacture discrete products, like automobiles, nor does it produce revenue from sponsors buying mention on information and entertainment services disseminated freely to the public.

Time sharing is a special kind of service industry based on the remote manipulation of data via certain combinations of technology—computers, communications, software. This becomes a viable business largely because the technology gets attractively inexpensive to use when its essential costs are shared.

If enough customers share the cost of a powerful computer and its programs, computing and data processing per se can be within the reach of an enormous number of organizations, both large and small.

The customers use keyboard or graphic terminals linked to a central computer by telephone lines. In effect, each customer believes that the computer operates exclusively in his behalf, for certain supervisory programs within the machine analyze the demands made by the various customers and interleave them so that the central computer works on one or another or another with almost lightning speed.

As a business, computer time sharing looked like a golden goose once the techniques of sharing a computer were demonstrated along about the middle 60s. It would be laying temporal golden eggs—the kind you sliced up into microseconds of operating time and billed for accordingly.

Some entrepreneurs, though not many, thought you didn't even have to buy a machine to get started. First you prospected for customers, then you gave a demonstration to the customer via telephone lines to the manufacturer on the kind of machine you were going to rent. After suitable juggling had been accomplished among the potential customers, a computer supplier, and the phone company, you hired some smart programmers and whipped them like galley slaves to have the software ready on time to meld the whole affair together in an operating system. At least that was the kind of precarious manipulation that launched one or two adventurous companies.

Technically creative people impelled most time-sharing companies. Many of them were started by programmers who were attracted by the idea of capitalizing on what they knew how to do with software—squeeze the most out of computer hardware. It doesn't take much scratching on the back of a discarded punched card to run out the figures: A time-sharing system, running prime shift 8 hours a day for, say, 10 customers at 100% utilization will generate "n" revenue. All one has to do is set the rental factor that will generate "n," then set up shop and become a very rich man.

This kind of arithmetic turns out to be technologically simplistic. A certain kind of queueing phenomenon exists in computer time-sharing: The facilities you need to provide reasonable service have to substantially exceed the average demand, or else the response time of the system seems too slow for everyone trying to use it. From a business point of view, you might find it uneconomical to operate at a smaller percentage of capacity in order to provide a nominal level of service.

Moreover, operations turn out to be at the mercy of customers' whims;

people call up the computer at very different times of the day and night. Because there is no steady-state kind of computer processing, being attuned to the system activities of the moment can lead into temptation. A computer system is clearly running under capacity—so why not sell that obviously additional capacity and make more money? After all, additional increments of computer time don't seem to cost a time-sharing company anything when they are as unused as a few empty seats after the curtain has gone up on a show.

The temptation becomes two-fold: First, one can sell those additional incremental minutes since they will be interleaved with other demands upon the computer. The only risk seems to be that the demands might all come at once and degrade service. Second, by taking that chance, the pricing strategy can be based on selling attractively inexpensive increments on the assumption of operating costs figured at near capacity of the computer system.

If enough companies put enough computer time on the market based on such an incremental pricing strategy, it doesn't take long before the cost competition has all of them selling at below real operating costs unless their computers do run at capacity which, you will remember, degrades service and drives customers away. Then, when people find their services aren't selling, they cut price. This iteration becomes disastrous when there is no distinction between time-sharing services except price.

So computer time sharing is not like a production business. There is not really a committed workload; certainly there is no backlog, for everything about the service operates in real time.

This principle probably wasn't well understood by the businessmen who flocked to start up or back time-sharing enterprises. Ever since time sharing had begun to look like a viable business along about 1965, it was written up as an industry of great promise. It readily attracted investment capital because the cost of getting into operation seemed relatively low.

The trouble was that many of the time-sharing companies, both new and old, had completely miscalculated their market, revenue projections, and capital start-up needs. Many of their troubles were technical. They thought they could get a computer, buy some software, have the phone company hook up data communications, then run the resultant system without any further understanding of how it worked.

But the systems often turned out to be curiously sensitive. Files would get wiped out, or simply get lost to float around somewhere inside the system. Sometimes the system would just crash for no apparent reason and fail to respond to anyone, not even its operators.

Whenever this happened too often, customers became discouraged. After all, few of them were skilled in computers—they just wanted service. But not many of the time-sharing companies had given much attention at first to helping the customer even get the most out of the services offered. There was little technical consultation to support him. If changes were made to the time-sharing system, documentation of the changes rarely followed fast enough to advise the customer that he would have to alter his own procedures in using the system.

When customers drifted away from fledgling time-sharing services that were under-capitalized to start with, it became harder for such companies to go back to the well for more money. A number of companies actually got into business on about $¾ million, which turned out to be roughly half a year of operating costs.

Capital for almost any kind of business began drying up by mid-1969. As the stock market began falling out of bed, investors began to sober up quickly about the truly great influx of time-sharing houses that appeared in late 1968. After approximately a half a year, the optimistic projections of these companies for high sales and low expenses began melting away.

Few of the companies had made any real investment in a marketing force. Many of them acquired a good peddler, knighted him with a title of vice president for marketing, paid him $25,000 per year plus stock options, and expected him to scare up customers for the computer waiting in the back room. Sometimes he was permitted to hire a couple of salesmen, although it wasn't clear that the peddler had the capability for management, too.

Meanwhile, whatever was being sold as a time-sharing service was changing drastically. Early in 1968 a salesman who knew a little about programming still could go into a scientific research or engineering establishment and offer raw computer time on a central machine via teletype link; the client could then do as he pleased with the system. But it became more difficult to sell this way because there were lots of other customers who examined an offer of a computer and asked: "But what do I do with it?"

There followed a great rush through 1968 and 1969 to produce libraries of computer application programs. This meant that a salesman was selling programs which a customer really bought by signing up for time-sharing service on which to run them. However, most programs weren't very difficult to generate in the first place and they were certainly easy to copy. Some time-sharing companies hadn't even taken security precautions against anyone dialing their computer and copying an entire program and related code straight out of the machine. Few programs were at all proprietary.

After two years the great flurry of programming activity hadn't produced much of a distinction between time-sharing services after all. Almost every service, large or small, offered the same computer languages and somewhat the same kinds of programs. This similarity of services continued to depress the time-sharing market.

The shakeout became apparent in the fall of 1969 when a rash of mergers broke out. A couple of time-sharing companies running similar systems would get together because they thought they'd save expense by not having duplicate computer centers. But if they didn't gain marketing strength or proprietary software by the consolidation, they still found themselves in trouble. Some of the giant time-sharing services began to retrench—and also raised their rates, which raises an important point:

A lot of underpricing of time-sharing services was based on the conclusion that the more clients there were, the less it would cost to maintain the central system they used. Time-sharing prices, it was reasoned, could be based on pro-rating the cost of the shared computer. But much of the cost of a machine and its operation is in the peripheral equipment. As machines get faster, there may be more central processing time to be shared, but printers and tape units aren't speeding up at the same rate. More peripheral devices and operators become needed as a system takes on additional customers.

Even if the customer load stays fairly constant, it may take as long as 3 or 4 years to bring a particular computer system to maturity in the sense that it becomes a cost-effective, useable service to customers via the kind of pro-gramming refinements that make the most of the computer logic to be shared. Thus, even a time-sharing service long established on an older machine may operate more economically than a service that first uses a new machine in the forefront of the state of the art.

Still another shakeout in the time-sharing industry may come fairly soon, when newer machines are brought to maturity by the services that acquire them. Then an otherwise perfectly good older machine will become a techno-logical anachronism: It will no longer be cost effective in the face of compe-tition. And the more capability that clever systems people can squeeze out of a machine, the more likely it may become that computer users will not only get time-sharing service out of it, but can buy a spectrum of other services, like batch-processing, from the very same machine.

Meanwhile, as the present shakeout continues, the strategies for survival and success may revolve around the ability of a time-sharing service to acquire many different proprietary data bases and develop lots of proprietary software that can tailor that data to each customer's needs. Such exclusivity is a good hedge against price erosion. Certainly it may cause a lot of special-ized data-gathering concerns and many time-sharing services to regard the prospects of marriage.

If all this seems to paint a black picture of the time-sharing industry's future, I don't mean it to. Time sharing holds significant promise as part of the spectrum of computer services on which there is money to be made. The companies encountering serious troubles can't really blame them on the present business environment, or on hard luck. There are various strategies to follow in entering a business based on time-sharing technology, and there are various ways to organize a company around the technology to turn it into a viable business.

Let us now see how one company was established to exploit the technology of computer time sharing, and what strategies that company used to gain its present relatively stable position in the midst of the current shakeout.

Later, we will look deep inside that company to examine the styles of its managers and how they organized the business and technical parts of the company, internally and externally, to meet the problems they encountered: we will see how these managers changed in their attitudes toward managerial and technical tasks at hand, and toward each other, too, as the business and technical parts of the organization came to further appreciate the equal importance of their roles.

First, let us place the company within the spectrum of the time-sharing industry:

Of the 140-odd companies now in the time-sharing business, a few giants are doing $10 million a year in revenue. Another 3 or 4 are doing between $5 and $10 million and a third group of perhaps 10-20 companies is doing $2 to $5 million. The rest of the 140-odd fall well below that.

Interactive Data Corporation, based in Waltham, Mass. and serving a national grid of customers, falls comfortably in that third group and its managers believe it will easily survive. They expect to be among the large independent companies with sales in the $50-100 million range sometime in the middle 1970s.

Part of the reasoning of the managers at Interactive Data is that their technical and business strategies are different: they regard time sharing as only one of several necessary ingredients to their business. Interactive Data Corporation, they say, is a data utility. First, rather than compete by selling just computer time to anyone for any purpose, as some other time-sharing companies do, the business strategy is applications-oriented. It is focussed sharply on the on-line data processing and information needs of particular industries or large corporations. Though other time-sharing companies are applications-oriented, few also have the proprietary data-bases involved in certain applications.

Interactive Data Corporation offers access, in big cities throughout the U.S., to large-scale reliable computers. In addition to the normal computer languages such as Fortran and Cobol, specialized languages such as FFL—the First Financial Language—are available which permit people with no computer experience to access, process, and screen large financial data-bases.

Interactive Data Corporation was formed in December 1968 out of two existing organizations that had already started operations on the East Coast. In New York City, White, Weld & Co., a major investment banking and institutional brokerage house, had been operating computer services for itself and its institutional customers to aid financial analysis and investment decision making. The service had been founded earlier by a technically educated, business-oriented general partner, Joseph J. Gal, now the Chairman of the Board and Chief Executive Officer of Interactive Data Corporation.

Meanwhile, in Boston, Mass., a young company called Computer Communications Center had just been founded by Jack A. Arnow to offer special purpose time-sharing services based on Arnow's previous experiences with one of IBM's most powerful computers, the 360/67. Arnow had a solidity and technical sophistication in the computer community dating back to the pioneering days of 1950.

I'll be telling more about the backgrounds of these and other key men as I come to them individually, for it will become apparent how they have been influenced in choosing their strategies and styles of management.

What brought Gal and Arnow together was the kind of natural affinity that comes in the expansion of new kinds of technology-based service operations: Businessmen and technologists need each other more than ever to sustain and grow a stable operation. When it becomes clear that the use of some technology at all does make a significant difference to customers, then one seeks to refine the role of the technology to maximize its potential for service and profit.

For refinements in the computer field, the common practice of leasing machines makes it somewhat simpler to go to larger of newer technology to keep up with demands on capacity or with new applications. However, the time sharing of computers among many users is a lot trickier affair technically than just having a computer of your own in the back room. To users, the machines are invisible, just as a power station or telephone central office is invisible; all a customer is really concerned with is service.

At White, Weld, it was becoming clear that a major investment would have to be made in new technology to provide more computer service. At the same time, the customers couldn't be expected to allow for growing pains in the transition to a more powerful computing machine. A virtuoso of that computer system would be needed and at the time, the most likely machines for time sharing were simply too new to have fostered many virtuosi.

If there was ever a master of an advanced time-sharing system at the time, it was Jack Arnow. He had been instrumental in the development of the IBM CP/CMS Time-Sharing System for the 360/67 computer at MIT's Lincoln Laboratory and was responsible for its implementation as a reliable time-sharing service there. This pioneering installation was so successful as to motivate Arnow to build a commercial service around such capabilities. He thought he could do it with the computing machine that had turned out for him to be a Stradivarius of time sharing.

Somehow Arnow must have begun to look a little like a Paganini. Gal had heard about his work at Lincoln Lab and sought to hire him, not knowing that Arnow had already formed his own company and was already in business —three days. He was planning to locate his computer system in Wellesley. At first they explored a contractual arrangement between the two companies, for White, Weld was already playing to its specialized audiences in the financial community. Then Gal and Arnow began to see how they could do even better in concert. The month before I had come on the scene, the White, Weld division and Arnow's fledgling company had become Interactive Data Corporation.

For almost any company, the first year or two of any start-up operation usually rushes by so fast that the ways an organization changes are rarely apparent. There is simply so much to do—to organize, to acquire a staff and make it mesh, to develop products, facilities, markets, customers—that nobody swept up in the hectic pace of it all can precisely detect change in passing, nor accurately recall it in retrospect.

There are changes in organization, in styles of management, in technical and marketing strategy, in the values put on people, or on activities, or on facilities. Some of the changes involve interpersonal relationships and ways of working together as organizations grow, and as problems grow. Not realizing the extent of change in an organization, or understanding the nature of various changes, a manager may still envision himself managing the company he helped initiate. Yet indeed the company may be significantly different in subtle, or at least unperceived ways.

To discover the changes at Interactive Data, I asked essentially the same questions of each manager early in 1969, and then again later in the year. The taped interviews were compared for evidences of change. As you will see subsequently, companies simply aren't made, they are born—and they are, for most part, grown from within by their managers.

In the next chapter, I shall present a more intimate view of the officers of Interactive Data Corporation. Though you'll be learning about their experiences and reactions in the context of a computer-based business, in the larger sense they are representative of most managers of new technology-based enterprises. Few of the patterns that emerged are peculiar to this organization, though in the business that it is in, it may turn out to have organized more efficiently than most other such companies.

If you've been through the excitement of a start-up, you'll recognize some of your own growing pains as you read how the managers saw the same problems but from their own—quite different—points of view, and what they tried to do about them. You'll also perceive a gradual unification of the company as its managers recognize the complementary strengths among them.

April 1970

6

How a Young Company Changes

Like most technology-based businesses, a time-sharing organiza-tion isn't simply born—it is grown from within by its managers. But the men who grow a company rarely perceive how they and their organization change during the tumultuous months follow-ing start-up.

Evan Herbert compares two slices of managerial life during one company's transmutation of computer technology into a viable business.

Mr. Herbert is a Senior Editor of *Innovation* Magazine.

There is probably no more fascinating challenge to the art of managing inno-vation than to take some apparently useful piece of technology and turn it into a viable business. As a case study of how this can go, I have chosen to explore the technical business strategies and the different management styles at work through the first year of life of the young computer time-sharing or-ganization, Interactive Data Corporation, described in the preceding chapter.

From that earlier discussion you know something of the current problems of time sharing, one of the newest of the high-technology industries to spring up in the last decade. In this chapter I shall show two slices of each manager's life at Interactive Data, the first at the beginning of 1969, soon after the new company started, and the second approximately 8 months later when, as you will see, many of the problems they had anticipated at the start had been replaced by quite new, and sometimes unsuspected ones.

I have chosen to let the attitudes about these problems emerge, for the most part, from the actual words of the managers I interviewed. The technical men turned out to be shrewd about the business, and the businessmen were not the least bit innocent about the technology.

Let me set the stage then by describing my initial visit to Interactive Data Corporation on a snowy January day.

My first reaction was one of surprise. It simply didn't feel like a brand-new company, or even a young one. The building, just off fabled Route 128, had the new, modern, solid-for-eternity feeling of a bank. Enthroned behind glass walls on the ground floor stood a 360/367 computer, then one of the few such machines in existence. In cost and capability, it would have an enormous effect on the strategies of the company.

If there was any hurried activity in the computer room that day, it was among the visiting telephone technicians; the computer operators moved about with the smooth nonchalance of professionals who had already mastered whatever crises could erupt.

The last felicitous touches were being put on the building interior—tasteful modern fixtures and furniture, extra long desktops and especially wide bookcases to accommodate the oversized computer print-outs and program manuals. It looked expensive and it was. And everywhere chattered computer terminals, as if everyone on the staff was connected to the computer by an umbilical cord—even the president.

Jack Arnow, President and Chief Technical Officer

I had expected Arnow to be as kinetic as the constantly flashing lights on the computer console downstairs. But he turned up affable and relaxed, like a father who had just married off the last of his daughters to hardy young millionaires.

Initially, Arnow nearly became an insurance actuary but one of his professors in MIT's electrical engineering department persuaded him not to waste his time. Arnow was already doing some numerical analysis for them and they wanted him to help out on some new, rather complex real-time systems, then being developed at MIT.

That was in 1950 and from this pioneering work emerged SAGE, the first application of digital computers on a grand scale to air defense. Development of SAGE had a significant effect on the technology of that era, and perhaps

Left to right: John Rutherfurd, Edward S. Greaves, Frank Belvin, John M. Thompson, Jack Arnow, Joseph J. Gal (seated).

even more profound an effect on the young men who managed these unprecedented technological developments.

You'll see that influence, for example, on Jack Arnow, who spent the first 7 years of his computer career on SAGE. He managed 200 people and was responsible for design, development and implementation of SAGE's first production program. Except for a year's leave of absence at MIT's School of Industrial Management, all of Arnow's technical management experience was in computer technology for the government at MIT and its Lincoln Laboratory.

What made him then enter the commercial computer world in 1968?

"I had a strong desire to do what people were saying couldn't be done outside that lab: time sharing on the 360/67. We'd done it at Lincoln, of course, but that was essentially a closed environment like so many government-funded laboratories and industrial research establishments. New computer system developments are locked in to the particular facilities at hand and aren't necessarily viable where the system would be in the hands of other users remote from those facilities and hardly familiar with them.

"Looking back, personal financial return didn't turn out to be the prime motivation so much as personal independence. I really wanted to test my hypothesis that what we were doing with that computer at Lincoln wasn't much different from the real world outside. If so, those techniques would have commercial viabilities."

Arnow's view of the business he entered is the delightfully straightforward approach of an experienced technical man. The objective of the game is to produce more revenue from a computer than he spends in support of that computer. As chief technical officer, he has committed himself to supply certain things out of the computer—programs, capacity, reliable performance. So long as those commitments are fulfilled, the board of directors will remain happy.

For the most part, though, fulfillment is visible only to the eyes of observers with particular and deep technical understanding of what goes on inside a time-sharing system. So there's a fair amount of trust buried in a technical business founded on advanced, complex technology and used in ways so sophisticated as to be beyond the comprehension of most of the company's directors.

"I believe that some of the success in management" observes Arnow, "comes with the recognition of each person knowing what he really doesn't understand. But this is balanced by confidence in the expertise of the other managers in their respective fields. Here one of my colleagues among the operating officers has more technical understanding of the internal computer system than I do, and three of them have technical understanding at least from the outside. But they have very strong talents in the financial and marketing areas where we technical men are limited. I couldn't sell a shoeshine."

The reference to trust and complementary strengths accounts for the dual way this company—like many others which operate on a high-technology base—is organized. Arnow is responsible for the technical operations and long-term technical developments. A vice president for computer systems reports to him, as does the manager of operations. Joe Gal, the chairman of the board, is chief executive officer and is responsible for daily corporate operations and long-term development of the company as a whole. Vice presidents for marketing and finance report to Gal. On paper the dual alliances looked neat to everybody that January.

The merger had become official the month before. The technical job, as Arnow viewed it, had been to have a reliable time-sharing system up and running. Arnow's acute appreciation for reliability—a vital factor in the time-sharing business—had been bred into him from his earliest days of pioneering real-time systems. Arnow was proud that nobody's files had been lost yet by the computer.

In fact, it seemed that his major problem was with the telephone company; no matter how well his computer system ran, he couldn't control the external links between it and the revenue-producing customers. This was a different sort of operating environment for Arnow because, at Lincoln, the telephone system interfaces with the time-sharing system were all internal. He was hoping then that these data communication difficulties were merely start-up problems that would be solved by the phone company which, after all, had some obligation as a supplier.

Just as the computer equipment was a technical resource to be brought to a measurable level of productivity, so was there another technical resource of similar concern to Arnow: software. A computer can't produce without programs. Nor can the system be attractive to customers unless the programs tailor it to their particular needs. Part of Interactive Data's overall business strategy was to make that tailoring elegant—elegant in the refined way that means computer time-sharing operations are so smooth that the technology doesn't get in the way of a customer's using the system as a tool.

To implement that strategy, a major investment was being made in creative programming. By what criteria did Arnow judge how well creativity is transformed into productivity—a difficult word to describe the intangible values of computer programs? Arnow focusses on the attractiveness of the system to the customers:

"The development of a new computer language that nobody is likely to use—to the degree that would produce substantial and steady revenue—I consider nonproductive, even though the accomplishment is good in its own right. A demonstration program that may draw new customers to the system is productive even though it may never generate direct revenue. Programmers, technical people, are a difficult breed in some ways but they are almost guaranteed to stay productive so long as they feel dedicated to particular goals of the company to which they themselves can relate. This takes an involvement with the organization, not just with the technology per se. Productivity means causing something to happen, directly or indirectly, that is going to produce revenue."

Arnow had been instrumental in all of the hiring, seeking people with the spectrum and capability to go across as wide a class of problems as the company had—capability to transfer skills, not a finite knowledge of technology.

"Some of the people I've brought in I knew only from afar, but I could tell a lot about them from the way they worked in their organization. That often gave me a better evaluation than bringing them in for a two-hour interview. I'd get a feeling they were simply my kind of people—not simply bright mavericks bubbling with ideas, but solid, unstarry-eyed performers who can follow through. Any imaginative manager easily can generate ideas that will keep 20 people busy for the next year; the problem is accomplishing something, rather than pursuing everything. I'm really concerned about concentrating our energies in just a few fruitful directions."

But meanwhile, how to keep a creative group from feeling their individual bright ideas were being stifled? Wouldn't Arnow turn them off by being arbitrary, mean, ornery?

"People have to understand your *reasons* for being arbitrary, mean and ornery so they can see what's behind the priorities for the execution of ideas. I've been hoping that this company could become like my group at Lincoln—a sort of benevolent dictatorial anarchy in which people could do anything they want, for which I would have a very wide range of tolerance because we'd all have agreed on the direction in which we were going as an organization."

This attitude stems, in part, from the creative nature of the technical tasks Arnow is responsible for. He concedes that he doesn't really know how to guide somebody through a complicated programming job. He can define what's wanted. He can talk about all the ways he knows of doing things. But all the managing, all the paper scheduling in the world is not going to get the job done. *People* are the means of production. Especially in computer programming, which involves a structured intellectual disciplining of particular technical resources, it can turn out that although 90% of the work will get done in the first few weeks, unanticipated problems may show up later on which must get solved before the program really works. This may take months.

Arnow traces his abhorrence of schedules and deadlines as management control techniques back to his days at Lincoln Lab where everything being done on SAGE was tortuously scheduled: "Everybody relied on that schedule because it really seemed to mean something—except that it was fiction in representing actual development of anything as creative as the programs done by people. I used to wonder whether the schedule was slipping in real time or faster than real time.

"We're in a little bit of an analog world here. If you do things right, something will always be developed by some time. Then you build on that, good as it is, even if it isn't the best. The steady progression is a better way of getting out creative work than deadlines."

Still, for a chief technical officer, there is the matter of keeping in touch with the system, which must come up like the sun every morning, and keeping in touch with his colleagues in the operation of the company. Via the computer terminal in his office, Arnow generates reports for himself which he finds useful in understanding what is happening in the system. Originally he had wanted to develop a program he could turn over to his operations people so that the computer would generate the kinds of reports he wants. But each day he would discover that there was something else he wanted to know about the system, so he decided to defer the development of his own management information program and simply use the terminal to explore quickly whatever he wanted to know about the system.

Clearly Arnow gets some satisfaction from being in touch with the system; more important, he is certain to suffer the same slings and arrows as his customers when the system is in trouble—except that he then understands what's wrong, and what has to be done to fix it.

How has such a tool affected Arnow's performance as a manager?

"I know that I would feel a complete sense of frustration without having access to this kind of information. I can tell, almost to the individual programmer or salesman, who should be doing more work. I know what the system is used for each day, and what it should have been used for. I have an understanding of what my people are doing, the progress they are making. It's not like being Big Brother in the sense of 1984; I don't really see what's going on except in a macroscopic way. I don't try to measure anything but I do keep in touch with accomplishment."

Keeping in touch is certainly not limited to the computer. In the initial stages of their growth, I observed that Arnow tried to maintain as unstructured a world as possible, especially in the programming area. When Interactive Data was started, no line and staff hierarchies were established because he wanted people to feel as free to get to him as he did in talking directly to them.

When the company was formed, the officers had agreed to have meetings every Tuseday and Friday. They almost made the first one. The next two were cancelled, and the third lasted one-half hour. They were all too busy putting out fires. Subsequently, the only formal meetings were ad hoc. But there were informal dinner meetings, which seem easier to hold.

How would Arnow go about making a risky decision? He held his finger up in the air and answered:

"That's the way all decisions really are made anyway. Almost every decision you make has a probability of being right. I still have the strange feeling it's more important to understand how you're going to extricate yourself from the consequences of a risky decision. The biggest problems arise when people spend much too long in deciding whether they should do something at all. You might just as well do it, and if it doesn't work out, get out of it quickly."

Those were Arnow's attitudes towards the circumstances surrounding him and his company early in 1969. About 8 months later, I began to ask him about the changes *he* had detected in his company. What was most noticeable to him was that its growth from about 30 to more than 90 people (it's around 135 as this goes to press) had made it less of a family organization. He simply didn't know everybody in the company any more and it was obvious that this bothered him a bit.

"When we were starting from scratch to build a vehicle with which to carry on our business, the problems and pressures were to maintain a good environment for growing and developing. We still maintain that environment —we're still growing a system and I guess we always will—but now we're more worried about servicing a daily operation, making sure each day's revenues are produced. In the beginning, we had taken over previous clients in the merger, but new sales hadn't reached a significant level. Now we're close to capacity on our original computer system and putting in a second system.

"We're devoting a lot more time to trying to maintain a very high level of operating reliability. We have some steady-state operation now in which I can anticipate that next month will be only a few per cent different from last month, instead of 30 to 40% different. But the change in everything about the company continues to be very large—not just in sales, or in technical performance, or what occupies our time as managers. We seem to live differently each month."

During this period of growth, it turned out that there weren't any really dangerous crises. But early in the summer, the rate of change throughout the company was simply so great that everything that wasn't well bolted down really started rattling.

For one thing, the telephone system through which the company provided service began functioning badly under the strain of its increased activity. Perhaps the difficulty was always there, but it didn't have a large impact on technical operations until the number of external users grew large. The programming staff in the building, with no data communications intervening, had always accommodated to the computer system in ways that outside customers could not. Now parts of the programming staff were being diverted at times to accommodating the computer system to the heavier demands upon it via the telephone system. This kind of firefighting, Arnow thought, had not detracted from the development of his programming people, but had detracted from growing more new ones.

As the customer load on the system grew, the programmers had been getting less and less access to the machine for development work during the day. By the summer of 1969, that problem had gotten solved by giving the programmers portable terminals to take home; they'd dial up the computer on the telephone in the evening when loads were generally lighter. And still they would show up for work during the day, too. Arrow felt he'd guessed right about the people he picked; they were still motivated by a sense of pride in what they were doing, and found a sense of accomplishment in programming or whatever, no matter what the pressures.

The creative programmers who originally develop a system rarely have the patience to deal repetitively with problems they've already mastered to their own satisfaction. The marketing department acquired programmers who enjoyed applications challenges rather than system development. So some of the technical operations Arnow was responsible for were germinated in a business-oriented department.

Eight months after my initial view of the company's operating life it was still just as hard as ever for Arnow to find time to communicate with the other operating managers all together. Once, during the summer, they had all gone away for a weekend of talk. They no longer go out to dinner together very often because a dinner meeting seems to cut into their working time just as much as a now regular Tuesday daytime meeting.

By the summer of 1969, some company-wide problems were beginning to emerge which looked touchy. Originally there was neither a personnel department nor a personnel committee. Arnow had been doing a bit of personnel work all along and regarded it as a real pain in the neck because it demanded more time than he was able to give to it. Now that the company was approaching a strength of 100 people doing many different kinds of work, he could no longer figure out a reasonable salary structure by the seat of his pants.

For example, the marketing department is sales-oriented whereas the technical department is organized according to technical ability. Yet both departments now had programmers. How to administer a reasonable set of salaries across the two departments was the problem. Arnow didn't like it, though he knew he would have to face up to it; but he simply wasn't willing to emulate the personnel policies of other companies which he thought were singularly unimaginative in the way they set up arbitrary grades, titles, and scales of pay. Emulation would have been the easy way out of it, but Arnow was afraid it would change the character of his company and he made sure that his concerns were reflected in the actual administration of personnel, which was handled by the new financial VP, Ed Greaves.

When I asked about a long-range plan for the company Arnow conceded that the longest-range plan was about a year ahead because he felt the company still hadn't passed the first hurdle, which he defined as when he could take a vacation. As he saw it, since January 1969 the company had passed several levels at which everything should have been smooth and fine—meaning steady-state operations he thought he could afford to leave for a while.

Was he still keeping his hands on the tiller via the computer terminal in his office—looking at the various aspects of what flowed through the computer?

"The worst feeling of frustration at this point is that I've gotten further away from the day-to-day technical operation than I would like to be. There are piles of papers on my desk that I must get at, but I would much rather be back keeping my finger on the pulse of the system. I haven't been on the computer system through this terminal in a couple of months now, though I go downstairs to the computer room every morning to take a look at all the listings of the day before. I write down all the problems, trials, and tribulations I see. Then I talk to various people about what happened and why. I also get a summary report on all system operations via computer printout that tells me much that I need to know about both internal and external system operations."

Though Arnow clearly enjoys being president, he was revealing his frustrations in moving more and more from the technical world to which he had been accustomed into a more pedestrian day-to-day operating world. He felt less in touch with the state of the art and as I left he was making tentative plans to attend a computer society meeting—something that he had not been able to do since he became president.

So that's how the technology of Interactive Data was being managed and grown by the most technically experienced officer of the company.

The chief executive officer also was a technical man to begin with, but came to Interactive Data after a number of years on Wall Street.

Joseph J. Gal, Chairman of the Board and Chief Executive Officer

In all the time I watched Gal in action he seemed to have stepped straight out of the classical annual report photograph—there is an aura of the-man-in-command that you expect to go along with the title.

Gal is an electrical engineer out of MIT, and a financial businessman out of Harvard Business School. When he was still a general partner at White, Weld, Gal firmly expected to make money with his Interactive Data Services division and after a while it became clear to him what the division needed in order to be even more successful.

In the first place, Gal recognized that time sharing was highly innovative and required even greater technical strengths to develop a unique, distinctive, and reliable service. He didn't just need "n" number of programmers—he needed the best programmers he could find. A lot of the brightest programmers have an instinct for the freedom to work on the unusual, which means that they seek out environments like MIT or Bell Labs or Lincoln. There's a delicate step in hiring such talent to work beyond the R&D environment—one must engender a feeling that their kind of unusual work could turn out to be even more constructive because of its obviously direct connection with revenue-production in a profit-making company.

To Gal in early 1969, the emphasis on superior people who could produce unique programs, who had strengths in managing technology, would really pay off later when the whole time-sharing field began to shake down:

"In the industry in which we have chosen to compete, there is such rapid growth right now that any hodge-podge of organizational structures and management techniques, unless they are latently bad, is bound to flourish for the moment. The more important strategy is to structure our company environment around the kind of people whose performance will provide a foundation for sustained growth."

One of Gal's decisions was to make the new facilities in Waltham look modern, successful and stable. This was taken, he explains, to create the intangible atmosphere associated with thoughtful management and a well-organized, efficient company. Gal felt this would help attract the superior technical people out of the not-for-profit woodwork and hold them; the effect on clients was the smaller consideration. And the decision didn't seem expensive when rent on the building was compared with monthly computer rental. Gal wanted the company to look like, feel like, and be a center of excellence.

When Gal sat down with me later in 1969, to explain how he thought things had gone in the past 8 months, his feelings seemed to match the expansion of his company:

"We have developed a number of proprietary products—especially in terms of the comprehensiveness of the data base available on-line, special financial languages, and a portfolio appraisal system called XPORT. If we have any problems at all now, they are more related to declining profitability among brokerage houses. However, this situation should be short-lived."

I wondered how the internal environment of the company had developed in the half year gone by with one person responsible for the business aspects and the other responsible for the technical aspects. In that period Gal and Arnow had begun to realize that such an arrangement is like guiding a sailboat through a storm with one person on the tiller and the other handling the sails—each needed to understand more of the other's immediate operating problems. Gal and Arnow were now in the middle of correcting this.

A secondary strategy for the company had been developed which was to supplement its initial idea of providing data utility services for the financial community organized around large data bases and special languages. The management committee (the operating officers) and the board of directors felt there was an interesting business potential in taking over in-house computer systems of small clients—those spending about $100 thousand to $250 thousand per year for complete service and support.

Since my last visit to the company, Gal and his managers also decided to organize around product managers for the creation, evaluation, and marketing of new products. He expected that product ideas would float upward in the company as the natural result of the internal reporting and communication procedures. He recognized some problems with that:

"People may not feed product pipelines so readily in the future if they fed the pipelines in the past and nothing much happened. We react slowly around here to some things because we are all very concerned with putting out quality products, and we want to know a lot about the potential of somebody's new idea—marketplace, cost, revenues—before committing important resources to it. Perhaps another reason for this slow reaction time is that no single manager is familiar with both major operating aspects of the company, marketing and technical, so it is difficult to decide intuitively on new product ideas."

I had noted that in the months gone by, formal hierarchies had emerged on the marketing side of his company, while the technical areas remained loose. Gal thought that the loose structure worked fine on the creative side, but that it was necessary to build little bridges to communicate. It wasn't so much a matter of keeping up with technical project responsibilities or even P&L. He needed those bridges to make trade-off decisions.

How did Gal think the arrival of the second computer had affected the company?

"Biting off a big chunk of expense like that was no small decision. We wanted to scale it to the way our business was flowing. But in a small new company, business fluctuates. So we kept the installation dates flexible for the second computer. When we put it in, it complicated everyone's life for a few weeks as we tried to hook it up in the special ways necessary for our business. Also, some additional programming work had to be done. In a couple of weeks the new system was running as rock solid as the first computer.

"We have now doubled our productive capacity, which may have added a millstone around the neck of the marketing VP. Overnight he has more to sell. But we have spread our clients out over both machines, and now we have some extra machine time our programmers can use for technical development during the day."

Theoretically the second computer might have been put in the geographic center of the nation's communications network. I wondered about electing to put it in Waltham, Mass., too.

"I don't know that it is clearly cheaper to put a national time-sharing computer in the middle of the country. The whole analysis is terribly complicated: There are trade-offs between computer and communications costs, both now and in the future, as the rates for communications services decrease. The more important consideration is to be able to provide reliable operations throughout the country while operating from a location where you can attract and hold good creative people and good managers. What's more, it costs about $1/4 million to install a computer site, so we'd just be increasing our expenses for operating and managing several discrete systems. We'd have to duplicate in each of these computer locations certain data which we now store centrally here."

Gal's continuing strategy had been to steer the company away from the trap of becoming primarily a technical activity with subsidiary supporting functions. He maintained a strong insistence that the company was essentially a business with production, financial, and marketing characteristics. Thus, unified goals would meld the technical and business activities and strategies. One such goal was to consciously develop marketable aspects of a time-sharing system, rather than to simply count on having some time-sharing technology to market somehow. Therefore, in such a technically-based industry, it seemed to me a bit unusual to find that his marketing man came from another field. But, over the long term, the company's products would be in many fields and professional management of large-scale marketing was more important to Gal than any detailed specific knowledge.

John M. Thompson, Vice President Marketing

In the computer field, which is just about 20 years old now, almost all the vice presidents you get to meet seem to have been born about the same time as the first computers. John Thompson wasn't really that young, but his youthful exuberance had carried him far in business since he left Cambridge University with a master's in chemistry in 1962. He wound up in what he calls a typical British situation—stuck in a lab and given nothing to do while he got the feel of Shell Oil Company by osmosis for a few years. He answered a newspaper ad in London and found himself in the USA working for the Celanese Corporation in international sales.

At the end of 4 years, Thompson was thoroughly enjoying himself running the Celanese commercial paint domestic sales organization out of Louisville, Kentucky, when a phone call came from Joe Gal, then running the Interactive Data Services division of White, Weld. The marketing strategy of the present Interactive Data Corporation is an outgrowth of that preceding association with White, Weld. Thompson and Gal had met socially in New York, where Thompson had first worked for Celanese. They had kept up acquaintances and Gal was on the phone making proposals to him to set up a marketing operation for computer services.

From paint to the high technology of time sharing?

It wasn't so much of a gamble, as Thompson recalled the first time I met him. A superb salesman to the core, he was masterfully trying to sell me on the connection, too:

"I was counting on the existence of certain common denominators throughout marketing which can be applied to all technology. There are customers who need to be sold in the first place, and then resold every day in the sense of being serviced and being reassured that they made the right decision to take this service. If one is bright enough he can learn what the customers' problems are.

"In the time-sharing business Gal wanted me to enter, many of the customers would be starting from scratch with me. More than anything else, the common denominators in running a marketing operation are attracting and motivating good people who have a sense of responsibility toward the customer. This is what selling is all about."

Thompson believed the marketing strategy for Gal's time-sharing services ought to be more strongly customer-oriented than technology-oriented. He could hire salesmen who knew all about the computer, which was, after all, simply a tool, or he could hire salesmen who could understand what the tool would be used for by the customers.

In 2 months, Thompson had studied 300 resumes and interviewed 50 people. He came to the conclusion that the kind of salesmen he was seeking simply weren't out looking. In the end, he reached out and hired two people he had known all along: He stole his stock-broker from Louisville and a mutual friend of theirs who worked for AT&T in Chicago. Then he got a couple of people through an employment agency he trusted—a banker and a clearly entrepreneurial chap who wanted to break out of his family's business.

Only the broker and the banker had any experience in the financial community. Not all of the new salesmen had sales experience. Two knew communications and data processing well. Each was given the system programming manual that a customer gets—one written in easy language to bridge the gap between that part of the computer technology a customer would have in his office and the kinds of problems he would seek to solve.

Somewhere between page 15 and page 50 of the manual, each new salesman overcame his fear of sitting at a computer terminal. In 2 months they were all using the computer to produce the cash projections for their territories. Most of them had developed a feel for what the system looked like from a potential customer's point of view; they would fit in with Thompson's strategy of worrying more about the customer than about the technology.

But Thompson couldn't ignore what was happening on the technical side of the company. If a system programmer wanted to make changes to the computer system, he did, and the customer eventually found out about it with some degree of surprise. There were rules and regulations to stop that sort of thing from happening; it was company policy to keep the customer informed at all times. Except that there hadn't been very strong representation from the marketing side—nobody said: "Dammit! Don't do that to my customers."

Thompson felt the role of marketing was not only to find the customers for the services provided by all that technology, but then to represent the customers as the technology constantly was changed. Not that technological changes—in hardware and software—were necessarily bad. Most of them were invisible to customers out on the ends of hundreds of miles of communications lines; they only felt the effects.

There had been a company function called customer education which was aimed at providing the customer with useful ways to look at the data base, nice ways to use it to pick out stocks. Thompson felt it more pertinent to make this a technical service function, one that would be just as good at teaching the customer how to use the system extremely well.

That wasn't so much a change in selling strategy as it was in servicing strategy. The salesmen would convince customers that the technology could be used in ways appropriate to their business, then technical services would help them continually to get the most out of the technology by providing precise, timely technical answers to every customer enquiry. Reorganization of the marketing department along those lines was completed a couple of months after the merger into Interactive Data Corporation took place.

A hot-line was installed, a phone that the customers knew had to ring only once and it would be picked up by an applications-oriented programmer whose terminal could display the problem.

Meanwhile, a couple of technical consultants were made available, like doctors on call, not only to discuss customers' problems over the phone, but to hurry to their assistance personally.

If all the customers were out at the ends of telephone lines, then why was there such an emphasis in early 1969 on the appearance and furnishing of the headquarters for the new company and its computers? That was part of the strategy, explained Thompson:

"We took a 15-year lease on the building for growth, of course. We were concentrating at first on markets in the financial community. Banks try to look like solid institutions, even if the assets are invisible. In the time-sharing business, you can't see the assets either; you look through the glass windows of the computer room and all you know is something must be going on.

"At first we thought we could bring the technology right into a prospect's office for demonstration on a portable terminal hooked to the central computer. But the potential customer usually would gather so many of his people for the demonstration that they couldn't all see what was going on by looking over the salesman's shoulder. They'd focus their attention on the terminal rather than on the sales message.

"So we dropped that idea and armed the salesmen with sample computer outputs. They could leave a portable terminal with the prospect later, but it was better to get interested customers to come into our offices first. Our potential customers would wind up visiting us here at headquarters and we wanted them to see an orderly professional atmosphere that would inspire confidence in our ability to serve them. Financial people tend more to prudence than extravagance; so do we and the modern, enduring appearance of our headquarters was the best way to communicate it."

A little more than half a year later, Thompson was marvelling at the change in the balance of what he had first thought would be problems and those things that actually turned out to be problems. There had been a lot of paper tigers and he was feeling relieved about those:

"I was worried about converting the subscribers on our first computer system at White, Weld to the 360/67 of Interactive Data Corporation. We kept far more clients than I expected. And we attracted far more new clients, too."

What was different for Thompson, though, was that his marketing organization had ballooned in 8 months from a dozen people to 38. He had anticipated that expansion, but not the difficulties of pulling it off in terms of the amount of time he would have to spend on it personally. Thompson had thought all along he would have time to act as sales manager, and still doesn't plan on hiring one because he expects to be able to grow him right out of his own sales force.

There was another difference in Thompson's marketing department over the half year. It had started out with no rules to live by; every single problem that came its way turned out to be an adventure in deciding the merits of a procedure to handle it. Now there were policies and procedures and reporting forms.

Thompson hadn't deliberately set out to establish such formality, but when it became apparent to him, he began to do it even faster because it made his life easier. He now concedes that his department is well structured, the most formalized in the company. He believes he tells his people much more of what is going on than any of his managerial colleagues because communications is so vital to marketing.

What's more, Thompson had to get started building up a middle management structure under him a lot sooner because there were sales offices in other parts of the country and they had to be managed locally. That would have been simple in a conventional company, or even another time-sharing company, but Thompson was convinced that the nature of the products of this data utility, the way they were brought to market, the very nature of the market—made Interactive Data a very specialized company. There was very little precedent each time they decided how to do something. So some fingers did get burned:

"One classic error, which ultimately didn't get us in too much trouble, came from our decision to go to potential customers in the financial community who were using competitive time-sharing services. We offered to convert their programs for them so they could run them on our system, which was obviously newer and better. But we hadn't figured that the only person who could convert it easily was the chap who wrote it in the first place. If one variable is incorrectly transmitted, you just can't work out the logic of the program. We hadn't made exhaustive enough inquiries to find out whether this scheme was technically sound, and perhaps our own technical people weren't firm enough in insisting we not even try."

If there was another classic mistake in handling the technology, it was one made by just about everyone in the field. Thompson conceded that he had thoroughly underestimated the difficulty of program documentation for the customers. While the source material comes out of the computer systems department, Thompson saw that he should have established the guidelines for his own technical writers much earlier. He felt that he was so concerned about giving people their heads that he failed to provide a framework for channeling their initiative.

Did the original strategy of investment in the building prove its worth?

"By now, some 6 to 8 months later, we have a useful comparison with the reactions of prospects who visited us in a more tatty environment on Nassau Street in New York where we thought such quarters were allowable for a struggling young organization. Clients, particularly in the financial community, want to feel they are dealing with a stable organization before they entrust their money to it. Moreover, if the product isn't really priced correctly, they don't think it's worth anything."

What had the data communications problem done to Thompson's marketing activities?

"If a customer tries to dial across the street and can't, he blames the phone company. If he has trouble dialing our computer, he doesn't know that it might be the phone company's fault; he says to hell with Interactive Data Corporation. When we had communications problems in New York, a major market for us, our sales people spent 80% of their time trying to hold onto the clients that were disturbed about the quality of data transmission. That left only 20% of their time for selling. By having our computer system people help the phone company spot transmission troubles, we've freed our salesmen to sell 80% of the time."

By now you will have noted data communication problems being mentioned by almost every operating manager. What was done about this supplier problem will be described later, though the solution was implemented by the computer systems department under Frank Belvin. The world Arnow seemed to miss was in the hands of the technical man he brought with him in the original merger that formed Interactive Data Corporation.

Frank Belvin, Vice President Computer Systems

Frank Belvin's title implies responsibility for the computing machine and all programming development. In a broad and fuzzy way, he is tied into the computer operations department, which runs the computer, but the operations manager is not really under his wing. Belvin reminds me of the classic Leyden jars encountered in Physics I; it's hard to imagine so much electric charge stored in a container of such apparent calm, yet somehow you sense that it's there just waiting to be triggered.

Though Belvin's father wanted him to learn accounting, Frank discovered he had real aptitudes in electrical engineering and math. After Georgia Tech and the Navy, he wound up with a master's in EE at MIT and a feeling that he ought to get into communications, which he did at Lincoln Laboratory. There he started to do some work on a computer and realized that it appealed to him more on a subjective level than anything ever had in the past.

He was hooked. A computer satisfied his passion for detail, his fascination with a deterministic situation. ("If there is something that isn't behaving quite right, I *know* there is a rational explanation and it will keep me awake until I find it.")

He asked to be transferred into Jack Arnow's group at Lincoln. Fortunately, the transfer occurred just about the time the laboratory needed to evaluate its computer facility. The 360/67 was under consideration, a machine developed in part as IBM's response to Lincoln's request for a proposal for a new computer facility with certain time-sharing capabilities, and Belvin had heavy responsibilities in the complex process of deciding whether to order the machine. Belvin had some previous experience with Project MAC, a pioneering development in time-sharing which MIT was doing for the Department of Defense. He began to share with Arnow what they both thought were some important concepts of what a good time-sharing system ought to be. Then Arnow ended up in the hospital for a while and Belvin noticed that, for all the apparent anarchy of the organizational structure at Lincoln, Jack Arnow's influence on computer developments had been significant. Arnow had really been managing his group very strongly in a rather invisible way. That leadership became conspicuous by its absence and nothing in the structure made it possible for somebody else to fill the gap easily.

When Arnow formed his own company, Belvin became one of the principals and later fitted neatly into the merger that produced Interactive Data Corporation.

As I first discussed his activities in early 1969, Belvin found the hardest part of his first real managerial position was getting accustomed to the idea that he is affected by what other people do, or don't do.

Looking back from his new position as vice president at Interactive Data Corporation, he could see that some of his frustrations at Lincoln stemmed from the lack of clearcut lines of authority. Everybody was a staff member and even the group leaders and associate leaders were just special kinds of staff members. Yet here was Belvin in early 1969, running a computer systems department and seemingly following the same course:

"There are no titles in my department. There are no clear-cut lines of who is responsible to whom, other than that the programmers know they all report to me. It might be madness to think of managing all these people. I'm pretty sure it's going to get out of hand some day, but I'm not yet certain how I'm going to let any structure come out of it. It's hardly clear that I ought to give managerial responsibility to the strongest person, or to the one who appears most capable technically, or whether it would be most effective to retain a committee-like affair with individual spots of authority for parts of the system."

He fancied that among his programmers in January 1969, he had acquired some bright mavericks in the nicest possible sense of the word:

"My main intention is to get a good person and then try to fit him in where he'll be creative and productive. If a problem appears, I'll try to fit him into a place where the problem will be minimized. I don't really look for team members; I speak of fitting only in the sense that there be no abrasive encounters. I suppose more than one real out-and-out maverick—one who insists on doing something all by himself—would represent a serious liability to me at this stage of the game because he could only work on longer range research; he would be a resource lost to me for the day-to-day operations of maintaining the system.

"So far, in the hiring I've done, I haven't thought about the potential of people as managers. What I'm looking for is an alter ego for the part of the system management I have been handling most closely: people with an attitude like mine—feeling that there's no such thing as an irrational event. They ought to be able to perceive problems from the wispiest indication that things are not quite right. I don't believe in gremlins."

Though Belvin thought the programmers could work effectively as a team, there were some mistakes he absolutely wouldn't let them make—mistakes that could cause the reliability of the system to be impaired. This reflected both Arnow's influence and Belvin's own experience at Lincoln.

So, within his own responsibilities at Interactive Data, Belvin was involved in instituting the policy of giving one person total dictatorial responsibility for the system for one week at a time. That person makes the dynamic decisions as to what changes are made to the going system that the customers are using. If it is running with a new programming system and it crashes, the operator notifies the computer's mother of the week (MOW) or duty officer.

Whoever has the duty is not only deeply knowledgeable about the system, but aware right up to the minute about all the newest modifications.

Meanwhile, on the marketing side, one applications programmer served each week as a programmer of the week (POW), a sort of duty officer to help any customer that had difficulties. He could talk to the customer by telephone: the MOW could be called in so that deeper troubles with the console system could also be detected and fixed. Thus flowed feedback to the development side to fix fundamental troubles that might be revealed.

In early 1969, the young computer system could still have its ups and downs—a manic depressive system—and Belvin was not about to leave it alone, for the operations group really did no programming system maintenance of any kind. Meanwhile, whatever went wrong within the system was documented in enormous sheafs of print-outs called computer dumps. Dumps were stacked all over Belvin's office that winter of the start-up and he was working his way through them just to know what happened, even though only to kick himself for not having seen various troubles coming sooner.

Meanwhile, Belvin had started his wish-list—a record of every comment, large or small, from programmers, operators, salesmen, and customers about what ought to be done about the system. It served him as both archive and prognosticator, and aid to longer-range planning.

One of the things that almost everybody must have wished was for something to be done about the telephone company. Belvin, with his sensitive antennae for problems that could grow into real troubles, was already aware by February of 1969 that company-wide effects would be felt from the earlier decision to locate in Waltham, then possibly the worst location for telephone service. But there are few places in the U.S. that are free of telephone problems these days. And they're not the kind of places that would attract programmers.

How did Belvin view his own situation that winter at this early stage of the new company's life?

"I'm sure that I'm at the bottom of my life as far as being caught in a time crisis. Finding myself in a true management position has proved to be the biggest problem I have had to face. I am slowly expanding my vision to beyond day-to-day problems. I really wanted to enjoy a much more leisurely pace of life, but what drives me is this powerful motivation to produce a good system, as impersonal as that may sound."

When I returned to Belvin's beehive roughly 8 months later, its programmer population had grown from 8 to 37. Belvin, a man who had been struggling to get out of the darkening shadows of demands on his time, had now emerged into the sunlight. Indeed, he was feeling expansive, seeking to hire even more programmers, to grow another 50%. How had he smoothed out his operations?

"I found, for my own sake, that I had to organize my department in a more meaningful way. I was becoming somewhat of a bottleneck, insisting that things flow through me. There were some pressures from other parts of the company to be able to get to programming people besides through me. They needed easier channels of access and the net result was that I needed avenues into which I could channel things, too.

"I decided to designate my most senior technical people group leaders. The selection had nothing to do with seniority in the company, just sheer technical experience and capability. I expect them to evolve from leaders into managers, first by managing technical work, then gradually taking over the personnel administration of their groups. I resent my being bogged down with paperwork. The only real advantage I see in my own administrative role is that it provides personal amplification of my efforts. Most of my group leaders are successfully using their new positions to amplify their own efforts. One of them is still coming to grips with reconciling his own loss of time to devote to the technical work because he is managing other people as well."

As time went on, Belvin gradually removed himself from the middle of these groups.

"If individuals in them would come to me I would refer them to the person in charge of the group. This has had its gentle effect in the long run of making it clear to the members of the group just how I view the responsibility. I still feel the paternal attitude I've had all along; I'd like to see my department remain a congenial kind of arrangement."

Perhaps Belvin was reflecting his feelings about the changes he had detected in the company as a whole. As it grew larger in the 8 months gone by, he noted, the company was becoming a bit harder, less personal. He observed that it was thinking bigger, that nobody could depend any longer on the historic origins of the company to predict what the managerial alliances would be on a given problem. Everywhere there were signs of a more mature outlook. And in Belvin's own department he thought the biggest change was that its individuals had become increasingly aware of what their responsibilities meant in terms of the operation of the entire company.

A dominant feature of this evolving cohesiveness was now the clearly channeled means of communications within his department and outside of it.

"I have a ridiculous calendar that has me having meetings all over the place, though it's really not ridiculous in terms of the results. About a third to half of my time is involved in formal meetings. Informal meetings take up another quarter of the week. I meet with some of my groups every week, some every other week. I like to have the group leader run the meeting. I don't make any attempt to stifle any thoughts I have, but I realize this creates problems simply because I am the boss."

One of Belvin's most important meetings comes each Friday afternoon. He assembles his group leaders, the outgoing POW, programmer of the week serving the marketing department, and the incoming POW. There follows a detailed rehash of what happened to the system during that week and what might be done to improve next week's operation. Belvin's people summarize the Friday meeting in a report generated on the computer and made available on Monday through the system to those who need the information.

This quick reporting scheme does not preclude the need for good documentation of the changes in the system, and for documentation of particular programs which are, in essence, new products. There is an overlap here with the marketing department where new products are concerned. Belvin's group helps develop the products. Belvin doesn't want to burden his people with the writing of documentation for these; indeed they may be too terse to communicate with customers. But he has established a formalism in each of his groups; they must maintain a book into which goes all correspondence, all technical ideas, all minutes of meetings, all changes to the system. Relevant additions to the contents of such books are routed within the company and the parts of the documentation that ought to get into the hands of customers are written up by technical writers in the marketing department.

How had he kept his computer systems department fluid in terms of interaction with professional colleagues elsewhere who were similarly advancing the state of the art?

"First, I have a little lend-lease arrangement which gives one of my programmers to a marketing region one day every two weeks. He spends a morning talking about problems, about innovations in the system to employees of the region, to salesmen and technical consultants. The afternoon is spent visiting clients already on the system to find out what they think of it and what we might do to fix it or improve it. Still, the really important innovations —those affecting the inside of the system—are essentially proprietary and we would not discuss them with anyone. Even with this sensitivity, I encourage my people to attend meetings of SHARE, the IBM users' group where they will meet technically very sharp professionals from other organizations. It's good for them, but I do spend some time with whoever will represent us talking over what I feel ought to remain proprietary."

By late 1969 a second 360/67 computer was added and Belvin enjoyed the challenges it afforded. He saw it as an opportunity for operating both a production system and an experimental system. There was some consideration of pretending that the second machine wasn't physically in the building so that the staff could get experience in operating a system that might well be located elsewhere in the country.

What had happened to Belvin's wish list?

"I'm still the editor of that list and it can be a real millstone. I haven't yet figured how to work it into the organization. I'm looking to the group leaders to take appropriate sections of it and make it dynamic. There are times now when I seem to use the wish list as a means of escape; when somebody proposes a modification to the system, I suggest we put it on the list. But I use it as a means of educating. We record how long, say, a modification ought to take—my people's judgment, my own judgment. Later, when we discuss what's happened and look at the wish list again, all of us turn out to be wrong, though I'm only usually only half as wrong as the least experienced. I think that everyone in the computer business in general makes a mistake of underestimating everything. But that's what happens when you're growing a new technology."

Edward S. Greavco, Vice President Finance

Most technical men hate to be bothered with the spectrum of corporate activity that involves financial affairs, and internal matters like personnel procedures and arranging contractual details with customers and suppliers. But there are business-oriented people who so thrive on it that they sometimes tend to shrug off high-technology based companies as just one more business to be managed.

The experiences of Interactive Data's vice president for Finance and its Secretary (and legal counsel) suggest that although the technology base of a new company may bubble with possibilities of glorious alchemy, the transmutation process requires a surprising level of effort on the business side.

When Interactive Data was formed as an independent corporation in December 1968, it became more apparent how much of the fiscal and administrative detail previously had been handled under the umbrella of White, Weld's own housekeeping operations. Nobody in the time-sharing operations there had really worried about who took care of such things. In the new company, it was expected that the Secretary, John Rutherford, would simply handle all financial and legal affairs; as the organization grew larger, specialists could be hired to carry the increased load.

However, the burdens that were there from the start led to an immediate search for a financial vice president.

By the time Ed Greaves came aboard, I had already come to know the other managers of Interactive Data and had learned to be amused by the uncanny physical resemblance most of them had to the archetype of their job functions, as you can see from the photograph of the company officers standing in the computer room. Thus, I was prepared for one more success by Central Casting. Greaves was no disappointment; he just looked like the kind of person to whom you turned over your checkbook and said: "From here on, keep me out of trouble."

Greaves, a graduate of Amherst and Harvard Business School, was fresh from Skelly Oil Company where, as assistant to the president, he had been involved in a number of diversification activities. For some time Greaves had been thinking about finding a start-up situation where he could get some equity and grow with the new company. In joining Interactive Data in February 1969, he figured it wouldn't be very different from the various businesses of Skelly Oil—just a matter of his applying analytical processes to the financial planning. That was what he was starting to do when we met soon after he became financial VP. Therefore I shall skip to a period about 7 months later to find out what happened.

Greaves hadn't really burned his fingers much in learning that the time-sharing business is so new that nobody anywhere could be expected to plan the start-up of a company based on this technology right down to the last decimal place. He had worked closely with the other managers to do a rigorous job of financial planning. So there weren't any real surprises for him in what happened; he was simply having new experiences.

The programming and marketing efforts began to seem gigantic to him, yet it was clear in this kind of business that they had to be above a certain critical size. From the standpoint of revenue potential, the 360/67 computer system had been an unknown:

"Nobody had ever operated a system like this before with such a product mix of program applications for so many customers. We're still trying to learn just how much revenue you can squeeze out of one machine. That kind of unknown had us wondering about the timing of putting in that expensive second machine."

Though Greaves had originally projected revenues throughout the year, he and his colleagues were surprised at the jump in revenue about the middle of 1969 when all the customers were finally transferred over from a smaller computer system to the 360. The level of usage had gone up and stayed that way for a couple of months. After quickly extrapolating this trend, it was decided to press for earlier delivery of the second machine. What the managers hadn't figured, and perhaps no other new time-sharing company did either, was the sheer attractiveness to clients of using a new computer. They began using it like it was free—until the first bills came in. The usage levelled off to a more steady state, which took the pressure off getting a second machine on any kind of accelerated schedule.

There were other technical activities to which Greaves had to accommodate his financial planning and control. The inefficiencies of the phone company had, at times, literally shut the doors to Interactive Data's customers. One day in New York, for example, customers were cut off from the computer in Massachusetts for half a working day. This cost the company irretrievable revenue, irretrievable in the sense that the empty seats of an airliner in flight are lost forever as revenue.

Greaves also was grappling with the decisions on what the policy ought to be toward deferral of the software expenditures in which the company was making a heavy investment; this is still a thorny issue in the computer industry.

In the interim since I had seen him, he had been working closely with Joe Gal on the launching of a subsidiary company which would get into the business of writing tailored systems for corporations and financial institutions to implement their own way of watching stock trends.

One change in Greaves was most apparent, even though he was a seasoned financial man. He had come to recognize that it may not be as easy as it once was to market futures based solely on relatively new technology. There is little proof that the features of particular kinds of time-sharing operations are worth specific amounts of money. So if, at some point, the company wants to undertake additional financing, it will probably get a better reception from organizations that really understand the sophisticated nature of what the company can do on a sustained basis.

John Rutherfurd, Secretary (and legal counsel)

If there was any change in John Rutherfurd over the months that I knew him, it was in his pleasure at being relieved of the financial activities when Greaves joined the company. Rutherfurd is a Princeton graduate with a law degree from Harvard, though I'm not sure this has anything to do with his being the most unflappable man I have ever met.

He had borne the administrative load in the early days of the new corporation and observes that Arnow and Belvin, his fellow officers who managed the technology, had a remarkable awareness of the kinds of administrative and procedural details that would be needed, perhaps more so than some of the people from the White, Weld umbrella. This turns out to be less surprising than it sounds, for the two men had come from Lincoln Laboratory where they were involved in running government contracts and in the supervision of a great number of hourly employees.

Rutherfurd was now deeply involved in the company's external relationships. In keeping with present practices in the time-sharing industry, a customer's legal relationships with a data utility are kept simple. When a customer signs the basic agreement it says essentially his only commitment to the company is to take a small amount of storage on the computer's magnetic disks. This is not purely capricious; the state of the art is such that a computer service or a data utility cannot be expected to have a good computer system or good data in every possible situation.

Most customers are experienced enough with computers to understand this. Even so, they appreciate that they are buying more than raw computer time. They expect to get programs and languages and in some cases particular data which they can manipulate with such facilities to make it more useful.

Rutherfurd spends little time with Belvin and the programming department, to which he can't contribute much, but much time with Gal and Thompson because of their involvement with customers. He also watches with interest the situation with a major supplier, the telephone company:

"From the legal standpoint, the phone company is on the defensive all over the country, especially for data communication and interconnections and attachments to their system. We're small at the moment and are not taking a fighting posture about service. It's hard to prove to either the phone company or to our customers just which of the three parties involved is responsible for what's gone wrong with the data flowing back and forth. We tell the customer we think things are going to get better because both we and the phone company are gaining more expertise."

Rutherfurd is probably right on that score, and for a reason that goes beyond that terse statement and back to a concern of those who managed technical operations. By summer of 1969, the company had come to despair of getting a fast fix from the phone company on data communications for two reasons. First, it took time to trace troubles in the computer system and make sure they were due to data communication problems and not the computer itself. Then it would take additional time for the phone company's trouble-shooters to arrive and rarely would they get the same men twice; each new trouble-shooter had to become familiar with the computer system and its particular relationships to Interactive Data's customer grid.

To keep the system serviceable, the company purchased an enormous amount of sophisticated test equipment to monitor the data communication terminals and lines. A man who knew intimately both the computer system as well as data communication techniques was assigned to watch for signs of various troubles and identify these to the phone company in anticipation of any disruption of customer service. This strategy of holding the phone company's hand had certainly worked. Outages in service because of data communication problems were being held to a minimum. What's more, the same strategy was being applied to other suppliers, for by the end of 1969, computer companies were finding it expensive to send service people to the rescue only to find the fault resided in the program or in the data communication lines. Unless a time-sharing company could prove that it was not responsible for the fault, it could incur expensive bills for service calls. So there was another pay-off to Gal's initial strategy of seeking the best technical people he could find: The company would not innocently assume financial responsibility for every fault in an enormous system of interdependent technology.

And in conclusion:

There really isn't any end to this story. As I described in the previous chapter, the shakeout in the time-sharing industry has begun even earlier than expected. From what I have told of how one company in the industry organized itself to carry on one kind of business in computing and time-sharing services, it seems likely that some of its strategies for managing that technology could ultimately prove highly successful.

It's too soon to predict just how successful. In the jargon of time-sharing operators, each business day begins when "the computer comes up"—meaning when it becomes available to its grid of far-flung customers. And each morning, for more than a year now, the computer system of Interactive Data Corporation has come up like the sun.

April 1970

7

A New Process Creates an Industry

When the laser was invented in 1960 it was deprecated as a solution in search of a problem. But one of those problems was found shortly afterwards when scientists at the University of Michigan in Ann Arbor succeeded in making holography or lenseless photography practical, and triggered a whole new industry—or rather a set of industries.

Around Ann Arbor holography has played a major role in shaping a dozen new high-technology companies. Elsewhere it was quickly seized upon by the aerospace, information-processing, and electronics industries, by biomedical companies, and—belatedly—by the chemical and photography industries.

Michael Wolff tells how it all began.

Mr. Wolff is Chief Editor of *Innovation* Magazine.

Had some savvy technological forecaster been plying his trade at the end of the 1940s he might well have set the photographic industry to developing a camera that could take pictures without lenses, pictures moreover that would be truly three dimensional. The possibility of doing this had already been demonstrated by Professor Dennis Gabor of London's Imperial College of Science and Technology, and his 1948 paper had been widely distributed.

But, if there really was such a fellow, his advice apparently went unheeded, and holography became the latest in a long line of inventions to come from outside the industry where one might normally have expected to find it.

Radar and lasers, electrical engineering and communications theory: This was the terrain from which holography eventually sprang. Correspondingly, the new discipline to which it belongs—coherent optics—is already taking on the dimensions of an industry that will affect much more than conventional photography. In the mid 1960s, only a few scientists at the University of Michigan were making holograms. Five years later nearly every large electronics, aerospace, and optics company was researching holography for applications ranging from information processing and storage to entertainment, medical research, and industrial testing. In addition, a couple of dozen new little companies were eagerly staking out some part of this territory for themselves. Annual research expenditures probably amount to no more than $30 or $40 million in the early 1970s, but some observers see in coherent optics a potential for growth that could conceivably approach that of the billion-dollar semiconductor industry.

In fact it is commonly said that holography is the greatest commercial discovery since Land's Polaroid film. But the two inventions are not quite the same, and the distinction that can be made between them, as far as a technical management looking for new ideas is concerned, is an extremely important one.

In brief, Polaroid film was a finished invention. From the moment of its discovery its ramifications were perfectly foreseeable. But exactly the opposite is true of holography. Since the time of its invention it has been full of surprises; and even today, twenty years later, the totality of its applications remains tantalizingly unforeseeable. Indeed this is its fascination to many researchers. They can still go into the laboratory, seize their new toy, and turn up the unexpected.

There is little argument that credit for the invention of holography belongs to Gabor, a brilliant inventor and scientist who belongs to that remarkable group born in Hungary around the turn of the century, and including Teller, Szent-Gyorgi, Szilard, Wigner, and Gabor's present boss, CBS Labs' inventor-president Peter Goldmark. Very briefly, Gabor's papers explained how the three-dimensional image of an object could be reconstructed by first recording the object's diffraction pattern and then illuminating this pattern with a well-collimated beam of monochromatic light.

Conventional photography simply records a flat, two-dimensional image by focusing the light reflected from the object onto a light-sensitive surface. A diffraction pattern, on the other hand, records the wavefronts that actually arrive from the object; in other words, it "freezes" the arriving electromagnetic radiation. The effect of this is to record not just an object's brightness but practically all the additional information about it, most notably the fore-and-aft location of all points in the scene. For this reason Gabor coined the term hologram, from the Greek roots for *whole* and *message*.

The result is striking. Viewing a hologram is like looking out of your window, in that as you move your head you can see around the people, trees, etc. Furthermore, a hologram can be cut in pieces, and each piece can be used to reconstruct the whole scene, though with reduced resolution and depth of field.

It should also be made clear that holography is by no means limited to recording light waves. Holograms can be made from acoustic waves, radio waves, electron waves, and x-rays. Indeed Gabor was led to holography by the dream of greatly improving the electron microscope by recording its output as a hologram. Interestingly enough, though holography is now practical for many applications, neither electron microscopy nor x-ray microscopy is yet among them.

Because of the low coherence of the light sources available to Gabor, his holograms could only be made for relatively simple objects and under special conditions. As a result, Gabor found no industrial interest in his microscope, and after a few years he virtually gave up holography as one of his "many failures." Thereafter, except for a few university researchers, holography lay dormant until the early 1960s. Then two young scientists at the University of Michigan, Emmett Leith and Juris Upatnieks, used the newly invented laser as a source of highly coherent light to make high-quality holograms. Their work marks the birth of practical holography.

Most accounts give the impression that it was the laser that was responsible for making holography practical. What really happened was far more complicated and, in terms of its effect on the industrial scene in Ann Arbor and elsewhere, more significant.

Four years after Gabor published his 1948 paper, Leith left Wayne State University with his master's degree in physics, and joined the staff of the University of Michigan's Willow Run Research Laboratories (now part of the University's Institute of Science and Technology). There a highly classified program to develop radar and other sensors for combat surveillance was underway.

As a member of the 20-man optics section of the Radar and Optics Lab, Leith was teamed with a high-level group of scientists and engineers who were trying to build what was probably to be the most sophisticated radar system then in existence. The problem they were attempting to solve was this: How to construct a special kind of airborne radar (a side-looking radar) when the resolution that was desired would be so high the antenna would be too heavy for the airplane to carry! The solution, which apparently originated with researchers at Goodyear Aircraft and the University of Illinois, was to use data processing techniques to combine the data obtained along the plane's path and thus simulate the effect of an extremely long antenna.

There are various electronic means of processing the radar returns to accomplish this, but the technique that eventually proved successful was an optical technique—one that in retropect turns out to be very close to holography. A group under the direction of Louis Cutrona, which included Leith, Weston Vivian, and several others, constructed what is known as an optical correlator. This is basically a filter capable of sorting out a collection of overlapping signals and displaying each one separately, as a sharp, well-defined point. This turns out to be closely akin to a hologram.

A radar using such a correlator was successfully demonstrated in 1957, and optical processing was subsequently applied to other radars as well as to automatic shape recognition systems.

In 1960 Leith was joined by Upatnieks, a tall, thin, 24-year-old from Latvia by way of Ohio where he received his EE degree from the University of Akron while doing optical engineering at Goodyear Aircraft. Upatnieks recalls joining the Radar and Optics Lab "because Cutrona sold me on the idea that optics and electrical engineering could go together—he was very excited about it." Upatnieks is regarded as an extremely meticulous experimentalist, and soon he and Leith "just started working together."

"One day," explains Leith in his careful, restrained way, "we came across Gabor's paper and simply out of curiosity decided to duplicate it. We did, and right then and there decided that this was a most fascinating field to be in. It's a fairly striking thing—you see this sharp image hanging in space with no object in the entire system that looks as if it could produce it. There's just an unintelligible smear on a transparency. So at odd moments, in a corner of the laboratory, we started investigating holography. In particular, we tried various ways of removing the twin image, which appeared in Gabor's holograms as identical to the reconstructed image but considerably out of focus. That was when we introduced the off-axis reference beam, whereby you essentially modulate the diffraction pattern by bringing in a second beam at an oblique angle via a prism or mirror. This idea of the off-axis reference beam, by which we were able to preserve very precisely all the phase and amplitude information in the scene, is very common in radar and communication systems. It worked very well and did improve the imagery.

"In 1962 the University purchased its first lasers and this made things quite a bit easier. Lasers were too expensive to borrow but they were built so that light emerged at both ends. We decided to use the light from one of these idle ends to see if we couldn't make holograms of solid three-dimensional objects rather than just of transparencies. There was a lot of failure along the way but finally by December 1963 we made some very good holograms of solid objects."

Gabor's holography had been limited to a special class of semi-transparent objects because of the extraneous twin image and other distortions which added a noisy background to the reconstruction. But Leith had recognized from his radar work that a two-beam interferometric process could produce a hologram in which the twin image and other interfering terms could be eliminated, thus making it possible to reconstruct objects of greater complexity.

Though Leith and Upatnieks had published several important papers between 1961 and 1964, it wasn't until the actual display of a hologram of a toy train at the April 1964 Optical Society of America meeting in Washington that the scientific community at large caught on to the drama of their work. Leith's eyes sparkle a bit as he explains that "here was a type of imagery radically different from any that had been observed before, and hundreds of people stood in line to see it. After the meeting we were deluged with phone calls and letters requesting information on how to make holograms."

At the University itself, the demonstration of three-dimensional holograms had created a good deal of excitement, partly because few people had really been aware of what Leith and Upatnieks were up to. Recalls one colleague: "In those days Emmett was a very introverted young man, and he never bothered to tell people much about what he was doing. In fact he had a famous four-drawer file cabinet stuffed with pads of yellow paper. Whenever he did some work he wrote it up on one of those pads and put it in the drawer. Time and time again someone would ask Emmett for help on a problem and they would find the answer in his drawer! I would guess he was privately well into the analogy between radar work and wavefront reconstruction by 1957, but I don't ever recall him making any announcement about it."

Though Leith's experiments had hitherto been restricted to his spare time, $10,000 was forthcoming as soon as it became apparent to the laboratory management what he had. This sum is remembered as being quite a bit for those days. Nevertheless "while there was some excitement about it, there perhaps wasn't as much as there should have been," observes William Brown, who was then head of the Radar and Optics Lab: "While it looked like an excellent piece of coherent optics work, from a technical standpoint we couldn't be sure whether we had an important scientific tool on our hands or just a curiosity. It took the subsequent work on interferometry and the use of holography for stress and strain analysis to reveal the great potential for applications that lay in the work."

The uncertainty over something that was essentially a small part of the laboratory's overall mission might conceivably have left off-axis holography in the public domain. It was apparently largely through the initiative of one man—Fred Llewellyn, then director of the Willow Run Laboratories—that the University approached Battelle Development Corporation, the invention development arm of Battelle Memorial Institute. The outcome of this overture was that the University assigned to BDC all "rights, title and interest" to the holography inventions made by Leith, Upatnieks, and several other members of the Radar and Optics Lab. In return, it is understood, 15% of any income received by BDC for the inventions will be passed back to the inventors via the University. The University will then split with BDC any royalties that remain after expenses. So far, however, there has been no income for any of the parties because no patents have been issued. "Lots of paper but no money," sighs one administrator.

BDC, which is believed to have spent at least a couple of hundred thousand dollars so far in developing the patents, has pursued an interesting tack. In seeking an appropriate licensee, they approached DuPont and Scientific Advances, which is a wholly owned subsidiary of Battelle. Scientific Advances was set up for the specific purpose of spurring the manufacture of risky technical products.

Why DuPont and not one of the big electronics or optics companies? Answers a BDC manager: "Probably for the same reason we approached Haloid with xerography. We were looking for someone with a highly technical base but who had no preconceived notions and wasn't heavily committed to any one way of doing things. Also DuPont had the experience in films that is so important for holography."

At DuPont the venture was looked at by a team from the Development Department which is charged with launching new businesses. The team was headed by Russell Peterson, a chemist who has since become governor of Delaware.

The upshot was that Scientific Advances and DuPont established a wholly owned subsidiary named Holotron. Located in Wilmington, Holotron now has exclusive rights to the University of Michigan inventions and is building up to a highly respected scientific organization to exploit the patents when they issue. Little is known about their plans, however, for unlike most new companies, Holotron seems to abhor publicity. Its president, Dan St. John, a PhD chemist from DuPont's Development Department, would discuss neither Holotron's work nor the size of its staff. However, industry observers interpret papers that have been published by Battelle researchers as signifying an interest by Holotron in pursuing the use of acoustic holography for nondestructive testing in industrial as well as medical applications.

Meanwhile, back at the Radar and Optics Lab . . . Leith, who by 1963 was heading the Lab's 20-man optics group, recalls that "Immediately after the laboratory demonstrations of 3-D holography that year, there was an explosion of activity. Everyone wanted to work on holography, and it wasn't too long before we were able to get enough money from government and industry to support eight or ten people in it."

During the next few years as research programs ranging from 1- to 20- or 30-man efforts started at practically every major technical company here and abroad, several discoveries were made in Leith's group that proved of considerable importance to industrial and scientific applications of holography. Among these:

> Robert Powell and Karl Stetson published on using hologram interferometry to measure extremely small movements, an application that pointed to the first industrial role for holography in nondestructive testing and strain analysis.

> Kenneth Haines and Percy Hildebrand applied interferometric holography to stress and strain analysis and showed how to make contour maps of complicated three-dimensional objects with holograms.

> Adam Kozma, Norman Massey, Albert Friesem, and several others published on such topics as high-density storage of holograms in crystals and thick-film emulsions, color holography, holography with ordinary incoherent light, and animated holograms. These and other areas were of course being pursued at other places as well and there was a time when papers were appearing in the journals at a rate of one a week.

Today while the number of researchers in Leith's group is still around twenty, these seven and most of the others from the 1964-1966 period have departed to pursue holography in the presumably greener pastures of private industry. Haines went to Holotron, his partner Hildeband to Battelle's Northwest laboratory in Seattle, Massey and several others to the new KMS Industries in Ann Arbor, and Powell now has his own consulting firm in holography.

Early in 1969, Kozma and three other members of the Radar and

Optics Lab left to form the nucleus of an Advanced Optics Center established in Ann Arbor by the Harris-Intertype Corp. "to develop new products for the printing, publishing, information handling, and electronic communication industries." The Center is also supposed to complement the advanced systems technology at Harris-Intertype's Radiation, Inc., subsidiary in Melbourne, Florida, a high-technology company whose president—to illustrate the inbredness of this new industry—is a former head of the University of Michigan Institute of Science and Technology and whose research director was a member of Cutrona's optical processing team. Consultant and chief scientist for the Center: Professor (of electrical engineering) Emmett Leith.

Leith devotes about 10% of his time to this new consulting job. The remainder is divided between administering the roughly $750,000 in contracts his lab gets from DOD, NASA, and private industry for holography and coherent optics work, and in doing research himself in hologram storage, interferometry, display systems, nondestructive testing and other areas.

Such activity notwithstanding, there is a certain wistfulness as Leith talks about the excitement of a few years ago and the people who have since left. While acknowledging that by supplying a goodly number of people skilled in a new discipline, his laboratory has more than adequately fulfilled the primary function of a university lab, he quietly admits "you always hate to see good people go."

Of course, as in any laboratory, top management keeps a weather eye on turnover and its implications for the "aging" that has beset some of the big mission-oriented government labs. So far little concern is professed, there being a feeling that enough graduate students are coming in from the University to maintain the lab's level of competence. "As long as you can keep at least one or two geniuses, you're OK," says one manager. To many people, Leith is one genius.

Regardless of its effect on the University, however, the research on coherent optics and other advanced technologies at the Willow Run Labs has had a dramatic effect on Ann Arbor itself. In 1953 the most technical enterprise in the region was very likely the Ford Motor Company's assembly plant in nearby Ypsilanti. Then Bendix set up a systems center under the direction of a University of Michigan electrical engineering professor. About five years later, the explosion of university spinoffs began, and this brought the number of high-technology companies around Ann Arbor close to seventy. Moreover these companies comprise nearly half the technically-oriented firms in the whole state. Heavily weighted toward the advanced technologies of information handling—optics, computer software and electronic systems—they have transformed a quiet university town into what local boosters like to call the Midwest's Route 128. Once lush cornfields now sprout industrial parks housing enterprises such as Crystal Optics, Data Optics, Daedelus Enterprises, Interface Systems, Photo-Technical Research, Photon Sources, Spuntech, Sycor, and so on.

Of these seventy, at least a dozen either began as a direct result of the Michigan optics work or depend on it for an important part of their product line. In what follows, I will try to give some flavor of their diversity by briefly describing five.

Conductron is the largest of the University spinoffs and because of the considerable financial success of its founder Keeve M. Siegel, it is the one visitors to Ann Arbor are usually told about first. Siegel is a well-known authority on radar scattering who in 1960 was a professor of electrical engineering at Michigan and head of the department's radiation laboratory. "That was a year," recalls Siegel, "when the University's difficulties with the state legislature were leading to rumors of payless paydays, and many of the professors I respected had left." Consequently Siegel, who had done considerable industrial consulting work, let it be known he was open to offers from private industry. Soon afterwards he had the backing from Paramount Pictues to start a company that would exploit the semiconducting properties of materials he had discovered while measuring the microwave properties of meteorites. It was the wife of a Paramount executive, says Siegel, who coined the name Conductron from this.

About a year later, however, Conductron got into radar in a big way. Cutrona, Vivian, and several others from the Radar and Optics Lab joined the company and a development contract was received for a synthetic-aperture side-looking radar.

Since then, Conductron has grown to a $73-million manufacturer of sophisticated radar and penetrations aids gear, avionics equipment, flight simulators and trainers, as well as optical signal processors for oil and gas prospecting and many other applications. While the work at its 63-acre site in Ann Arbor and at its other divisions in Missouri and California is still largely military, the company is using its background in coherent optics (where it has some twenty-five people working) as a base upon which to build a commercial product line in holography. Conductron (like at least half a dozen little entrepreneurs) sees a big market in advertising and educational displays, and hopes to introduce some products.

Siegel no longer heads the company however. In January 1966, growth appeared limited by an inability to make the transition from a strictly R&D firm to a production firm, and Conductron merged with McDonnell Aircraft. Several months later in what he calls "an honest difference of opinion between Mr. McDonnell and myself concerning the future growth of Conductron," Siegel resigned. A few weeks later, with $2 million from the sale of his Conductron stock, Siegel launched KMS Industries. Joining him at KMS were Cutrona, Vivian (then just back from two years in the U.S. Congress), and several others from Conductron as well as—once again—the seminal Radar and Optics Lab.

While KMS started out in military research, it rapidly expanded into product areas as far removed as electro-chemical machining equipment, color processing kits for amateur photographers, educational instruction systems, multipurpose fasteners, books, toys and games. During its first fifteen months KMS acquired fourteen companies and some 1,000 employees, and went public. At the end of 1968, it reported sales of $52 million and by 1970 it had about 3,000 employees in thirteen states.

In explaining this product potpourri to the Detroit Security Analysts, Siegel put it essentially this way: "Our company believes that the future belongs to science, engineering, and advanced technology. While other companies lead with their dollars, we lead with our scientists. Therefore, we will go into only those product areas where our scientific talent excels and where its efforts can be expected to yield an improvement in sales amounting to at least 20% a year. Anything less would make us a routine company and be a waste of this talent."

As a result, KMS Industries is organized into what it sees as five major growth areas: optics, electronic systems, leisure time activities, education, and industrial technology. Insofar as optics goes, Siegel boasts that he has the best coherent optics group in the commercial world. Like his neighbors a few miles away at Harris-Intertype he sees printing and education as specific growth industries that are obvious outlets for this new technology. He is somewhat bolder than they might be, however, in claiming that "ten years from now we'll be looking at 3D TV and movies; two-dimensional viewing will be as archaic as the quill pen."

While entertainment could clearly be a mammoth growth industry for holography, the severe technical problems which need to be overcome before holographic movies and TV become commercially feasible make most holography experts hesitant about predicting when they might arrive. However, it seems clear that KMS means to be on the scene when they do arrive.

Meanwhile the firm is pushing the use of optical correlation and holographic techniques for fingerprint and other kinds of identification. Like quite a few other companies, KMS sees this as a promising application and one that could be potentially important for credit card identification and other security systems.

While companies like KMS Industries and Conductron are developing broad systems capabilities in coherent optics, several specialty companies are also springing up. One of the most talked about in Ann Arbor is GCO, Inc., started in 1966 by Ralph Grant for the purpose of applying holography to nondestructive testing. Early backers included Uniroyal, Ford Motor Company and Rohr Corporation.

Grant, like Siegel and Leith, was another Michigan faculty man from the Electrical Engineering Department with considerable industrial consulting experience. He conveys a brisk, professorial manner as he relates how he worked on the same radar contract with Leith at the University of Michigan's Institute of Science and Technology. While Leith worked on some of the holographic concepts, Grant worked on the development of solid-state light amplifiers. Says Grant: "During the very early stages of the development of holography, Karl Stetson, one of Leith's colleagues, brought to my attention that the holographic experiment he was conducting was extremely sensitive to vibration. That gave me the idea for using holography in an underwater sound-detector device. This concept was funded by the U.S. Navy and led to further Navy work at GCO on sonar transducers. The sonar transducer work, in turn, gave me the idea for using holography in the totally unrelated field of nondestructive testing. Today with holographic interferometry we can easily measure small subsurface defects, some of which are undetectable by any other nondestructive method. Deformation or displacement over a surface, resulting from mild stress, can be measured to within several microinches."

Out of this principle has come a line of four machines. One, which some tire companies have bought for $75,000, is a holographic tire analyzer, which reveals such subsurface defects as inner-ply separations and displays internal structure including splices. Another, a sandwich-structure analyzer, spots unbonded regions between the honeycomb cores of aircraft sandwich panels and their outer and inner skins. It sells for between $76,000 and $250,000.

Around Ann Arbor, GCO is spoken of with envy as well as pride, for it has grown to sixty people and $1.5 million annual sales, much to the admitted surprise of some scientists. Ten holographic nondestructive testing machines had already been put to work in the automotive and aircraft industries by the end of 1969, and Grant expected to have sold twenty-five more by the end of 1970.

Such growth naturally invites questions about competition, but Grant professes little concern: "Holography is a rapidly expanding field. Even in its infancy there is much more opportunity in the overall holographic field than there are companies to compete. We specialize in only one segment of the developing market, that of developing and manufacturing systems for holographic nondestructive testing. Moreover, the field we've chosen is technically rigorous and requires a great deal of expertise beyond holographic competence. It is one thing to create holographic images of chess men, or even interferometric holograms of small objects in a quiet lab. That's kindergarten compared to the real world of holography. It's a completely different ball game to build a 15-ton machine that operates reliably day after day in an aircraft factory, where there's noise and vibration, testing critical parts more reliably, cheaper, and faster than any other method. And that's the name of the game for us—not just making pretty three-dimensional displays but solving realistic nondestructive testing problems practically, quickly and economically—particularly where safety is involved."

Similar views are held by Don Gillespie, the 35-year-old, crewcut president of nearby Jodon Engineering who maintains that another holographic or laser company can only help the rest in town. "But," he warns, "this is not an easy game to play—it takes some smarts."

To Don Gillespie, having the smarts means being able to build complex optical instruments from the ground up, a skill he learned from growing up professionally with the laser. Gillespie was still an undergraduate EE student at Flint Junior College when, in 1956, he became a technician in the Willow Run laboratory of Chihiro Kikuchi. Kikuchi had been the first person to demonstrate maser action in ruby, and his laboratory was noted for its considerable competence in this new field.

At that same time another student from Flint was also working for Kikuchi, an undergraduate physicist with entrepreneurial ambitions, Lloyd Cross. In 1960, Cross, together with Gillespie and another friend, started an after-hours company named Trion Instruments.

Trion was set up to manufacture ruby maser radiometers, but later that year Ted Maiman of Hughes Aircraft announced the first laser and Trion was soon in a new business. On the basis of the experience with ruby at Willow Run, Trion received a $4500 purchase order from Texas Instruments for what if it wasn't the first commercial laser was pretty close to it. The order called for 45-day delivery and Cross remembers it as a frantic time. "Many people didn't even believe in lasers then, and we had no plant, no equipment and no parts. We just ordered the ruby, the capacitors, and everything else, and put the whole thing together. Four of us worked day and night and were maybe a week late."

Gillespie, however, had been doubtful about how well ruby lasers would sell and by this time had sold out his 12% interest in Trion, moving on to supervise the graduate student laboratory for Kikuchi who was now a professor in the University's nuclear engineering department. In 1961 the first continuous-wave, helium-neon laser was invented at Bell Telephone Laboratories (where, the story goes, the ruby laser would also have been "invented" had not some experimental work indicated, erroneously as it turned out, a very low quantum efficiency for ruby and thus discouraged a push to actually build a laser). Immediately, Kikuchi and Gillespie started building gas lasers and soon Gillespie had convinced himself that these would be far more important than pulsed ruby lasers. As a result Don Gillespie teamed up with his older brother John to form Jodon Engineering and manufacture gas lasers in the basement of his home.

From the beginning the Gillespies have followed a bootstrap philosophy of doing everything themselves. Thus, Don taught himself not only accounting, but glass blowing, refrigeration engineering, and all the other skills necessary to build a production shop with an air conditioned, electrostatically filtered clean room, a high temperature glass oven, and a glass working facility—everything necessary to build lasers from the raw glass and chemicals. At about the same time, the brothers added a third stockholder as manager of production design engineering, Gary Vander Haagen who was then completing his graduate work at the University. With this background as one of the few firms in the country able to build the heart of a holographic system, Jodon was able to move into holography as soon as Don got excited about it from seeing Leith's holograms reproduced.

Don believes he was the first to advertise holograms, recalling that when he bought a $115 ad in an electronics tradepaper he wondered if he would even get enough sales to cover its cost. "However, the paper was good enough to let me pay over a three month period and I sold about half a dozen holograms at $100 each. Next I got a list of the forty-odd companies then in the laser business and batted out a letter to each that sold some more. And then it just caught on and went."

So did Jodon Engineering. Today it sells not only gas lasers but a line of about twenty holographic instruments—attenuators, filters, power meters, etc. The Gillespies are out of the basement now; they have 6,000 square feet on Enterprise Drive, next door to a cluster of other one-story metal and masonry buildings which house Daedelus Enterprises, Balance Technology, Information Instruments, and several other young and growing companies.

Jodon brings out a new product roughly every three months, and John and Don Gillespie speak with confidence of having the smarts to keep ahead of any big company that might decide to start building the instruments needed by this new industry. Claiming to be one of the country's three largest suppliers of precision holographic instruments, with customers here and abroad, Don Gillespie feels that being able to build everything himself—and being located in a region he believes will be to coherent optics what Palo Alto and Boston are to physics and electronics—will keep him well ahead of the pack.

While companies like Jodon and GCO Inc. are being built to serve the technologically sophisticated, Lloyd Cross (who, you'll remember, founded Trion Instruments, along with Don Gillespie) has followed an odyssey that has brought him to an opposite position. Unlike most of his colleagues in the Ann Arbor optics community, Cross is a longish-haired fellow as interested in art and music as in physics. For this reason he is presently trying to make holography as simple and as accessible as photography. The story runs like this:

For several months after selling its first laser to Texas Instruments, Trion Instruments had the commercial laser market practically to itself and sales reached about $500,000. However, as so often happens, the initial capitalization was so small that a financial squeeze soon dictated a merger with a larger firm. In 1962, Trion became a division of Lear-Siegler, a diversified manufacturer of military and industrial systems. Cross managed the division and continued his research on pulsed lasers until 1967 when he left to join KMS Industries. There he worked on a problem that is still of major importance to holography—the development of pulsed lasers with characteristics that allow making holograms whose quality is as good as those normally made with helium-neon cw lasers.

Pulsed lasers are desirable because they don't require the stable conditions one must have when using cw lasers. Though helium-neon cw lasers presently cost a couple of hundred dollars, the giant granite slabs often used for stabilization run $2000 or more.

In the course of this work Cross needed to build a type of beam splitter where the reflective coating is deposited on an extremely thin substrate, ideally one thinner than 1/10 the wavelength of light. This device turned out to be so extremely sensitive to sound vibrations that the slightest noise would cause a laser beam hitting the substrate to be distorted. It was while watching this "visualization of sound" that Cross realized he had the answer to something he had wanted to do for many years—make music a visible as well as audible art form. "I found it was possible to make beautiful, fantastic displays with extremely high contrast, and when I showed them to other people, they got excited too."

Cross was unable to interest the optics group at KMS in pursuing the entertainment aspects of his discovery and so he had to obtain backing from several local individuals and start his second company, Sonovision. Sonovision is aggressively pushing the development of machines for throwing "a symphony of ellipses" onto the walls of coffee houses, theaters, and discotheques.

Besides this, Cross' long-held interest in having the ordinary public better appreciate art has led him into a strictly holographic venture. He and his wife Linda run an art gallery in downtown Ann Arbor, and with a lively group of young artists, are establishing a commercial studio to explore holography as an art form. An essential element of this plan is to take holography out of the laboratory by developing equipment simple enough that anybody can make a hologram cheaply and quickly. Cross claims that he and his associates, all of whom are without prior scientific training, have already built an inexpensive stable platform and associated equipment simple enough that any artist can easily make a hologram once he has a cw laser. During the early 1970s, the group plans to lecture at art schools and elsewhere on the use of this equipment, which they would then sell. "Within a year or so," exclaims Cross, "I think there will be hundreds of little holographic studios all over the country with people exploring holography the way photography was explored. Commercial holography is now where photography was in the mid-19th century, and the next step will be to develop a simple, cheap pulsed laser—this will do for popular holography what the flash bulb did for photography."

Quite a few people compare the present stage in holography's development with photography, which is why it is interesting to contrast Cross' optimism with the cautious remarks of a Polaroid executive:

"I'm not sure holography will be of any great significance to photography for several years yet. Insofar as ordinary 3D photography goes, there are simpler ways of achieving the same effect. But maybe I'm just sounding like the traditional fat cat. However, if you consider information storage and retrieval then holography has many interesting possibilities. But this hasn't been our field and we'll simply continue to keep abreast of developments."

Keeping abreast of developments seems to be the position of most photography companies, who currently see more profit in supplying film for holographers than in pursuing actual applications of holography. Why this position and not that of the pioneer?

Kendall Preston, an optics research manager for Perkin-Elmer, expresses a common opinion when he says, "The understanding you need to be really effective in holography must come from an understanding of electromagnetic radiation, not silver halide emulsions. Therefore, you need a concentration of people whose background lies in electric field theory, and this invariably means electrical engineers and physicists rather than chemists." Preston also points out that the heavy security wraps on the Michigan work could not have helped but inhibit the spread of information to the photography companies. "Even at Perkin-Elmer, it wasn't until 1960 that we learned that lenses we were building for Michigan were to be used in a holographic system," he says.

Then too, some people feel that information on holography might still have been disseminated more rapidly had it not been for a bitter controversy that existed at Michigan between Leith's group at the Radar Laboratory and George Stroke in the Electrical Engineering Department. Stroke is a widely known authority on diffraction gratings who came to the University from MIT as a professor of electrical engineering and head of a new Electro-Optical Sciences department.

Stroke and some of his students began working on holography in the same building with Leith—the "neo-Aztec" Institute of Science and Technology building which Ann Arborites universally refer to as Fort Apache. Presumably this is because of its fortress-like appearance, but an outsider might infer that it is also due to the verbal fusillade that accompanied a series of disputes over such matters as the credit for having been the first to suggest the use of a laser to make holograms. These snowballed over the years into a nasty series of interdepartmental squabbles, memos to University administrators, alleged slurs at scientific meetings, and letters to journal editors.

The importance of this—other than to show that holographers are no less human than other scientists—is that it inhibited some researchers, both here and in Europe, in their discussions with visitors. Though the controversy has simmered down now that the disputants are separated (Professor Stroke has moved to the State University of New York at Stony Brook), a number of widely traveled scientists hold that this lack of communication put a damper—for a while anyway—on the rate of growth of the field, in academe as well as in industry.

November 1969

8

A New Process Gropes for a Market

Perhaps "creating a market" is the way to describe the state of holography now—better yet, "creating markets."

With the entry of the big aerospace, communications, and imaging industries, and with big-time research, many possible applications are being turned up. Twenty-five new industries might emerge around this one intriguing tool and even then barely rub elbows.

Nilo Lindgren describes just some of the possibilities.

Mr. Lindgren is a Senior Editor of *Innovation* Magazine.

After Leith and Upatnieks displayed their toy train hologram at the 1964 meeting of the Optical Society of America, a lot of people in industry and in the universities began "playing" with holography. In some cases, they played with it out of sheer fascination with a new technique without worrying about any particular directions. Directions, they knew, would come later.

Those who had lasers available just wanted to learn how to make holograms and to explore what might possibly be done with them.

Some people recognized instantly that here was something significant, and they jumped in.

One result of all this activity was a spate of technical papers that suddenly appeared in 1965. What this evidenced was that researchers in many different laboratories had begun to discover that holography held more than met the eye. It was not, as some imagined, a kind of trick photography, just a minor sideline of laser physics. Those who had been trained in communications and electromagnetic wave theory sensed the potentials, but even they had surprises in store for them.

Once their researchers had seized on holography, the big question confronting the entrepreneurs—in the industrial companies, both big and small, and in the university environment as well—was how to find profitable applications that were relevant to their existing businesses and interests.

How the really big existing industries—those in aerospace, communications, computers, photography—reacted to this new invention is a story in itself.

To these giants, holography is just one more drop in the technological bucket. On the one hand, the giant can afford to keep small research groups going on many new technologies, so that in case something really interesting turns up, the company will have a "capability" in the field. But what one aerospace giant, for instance, does just to keep abreast of a particular technical field would create a sizable corporation in Ann Arbor. Thus, once the giants get into any act, they tend to dominate.

Bell Labs and IBM, for example, have become interested in holography largely for its potentials in optical information processing and for information storage and retrieval. The CBS's and the RCA's think of the mass entertainment industry. The photographic and xerographic industries tend to think of the market for the materials that holograms are printed on (there is potentially a market for zillions of miles of film). The semiconductor industry looks for new ways of printing circuits for higher yields. Manufacturing industries look for methods of testing and quality control. The western aerospace industries (and the university research departments with which they tend to become allied) think of holography in terms of applicability to their R&D. They look for machines that can test aerospace parts—nose cones, wings and fuselages, rocket engines—more adequately. Moreover, they look more than longingly on supersophisticated surveillance systems for the military.

But the big industrial companies face another kind of problem with holography. Despite the research thus far, holography, with a few exceptions, is just emerging from the early, exploratory stage of research and coming to the first applications stage. It doesn't yet pay for itself. What holography needs now, to speed up *all* of its potential applications, is a really big money-maker.

To spur and to sustain really sizable R&D efforts in holography, the giants need the prospect of a huge market of some kind. They need something that, in effect, puts holography somehow into every home or every plant. If such a market can be found, then holography can begin to pay for itself, and a basic present limitation would be removed. For at this time, as all the researchers assert, the field is limited only by the number of people working in it and by their imagination.

Some of the directions of this search for markets, for profitable applications, are what we shall treat now:

At this time, things have crystallized enough to reveal that there are several major thrusts or channels of holographic activity.

One is a very technical direction, embracing the automatic character-recognition systems used in conjunction with computers and optical data processing, as in reading books or printed data into computer storage. Holographic information systems could be used in credit card verification, in automatic teaching machines, in biomedical diagnostics, in fingerprint analysis, in mechanized surveillance systems, and so on.

A second technical direction involves the use of holography as an industrial and research tool. Such uses would include nondestructive testing, the visualization and accurate measurement of strain induced in objects under stress, the measurement of shock wave phenomena, contour mapping, photogrammetry, new methods of quality control, and the control of various electrochemical processes.

A third major thrust may be seen in what one might call the pictorial, artistic, or entertainment fields. These might include, for instance, art works, unusual displays (for education, advertising, etc.), three-dimensional television, and movies.

Time will undoubtedly add many more potential applications to this rough list, and perhaps even major new directions as yet unexplored. Because things can be done with holography (and with coherent light) that had previously been unthinkable, the field of possible and unique applications is far from being saturated.

When one compares, for instance, the potentialities of a new industry growing out of holography with what happened to the transistor, one factor becomes strikingly evident. When the transistor was invented (also in 1947-1948), it was virtually a one-to-one replacement for the vacuum tube in its functions while being packaged in a smaller, inherently more reliable, less power-consuming, and clearly cheaper unit. It quickly seized its natural market and, for many applications, made the vacuum tube a thing of history. And then it went on evolving.

Holography, on the other hand, is a genuinely new tool, an innovation that must create and find new markets. It is not photography, it is not just 3D, and it will not be one industry. It will conceivably be many, all barely rubbing elbows with one another.

It is for reasons like this that no reliable dollar figures can be projected at this time for the eventual size of the holography industry.

Although many techniques and many new technologies have been applied in computer memories, in data storage, processing and retrieval, holography offers some powerful, unique, and clearly competitive capabilities. In fact, many informed people predict that information storage and processing, including many kinds of pattern-recognition tasks, may eventually constitute the lion's share of the holography industry.

In brief, what holography offers to the computer and automatic pattern-recognition fields is a really high-density, high-capacity memory with a very fast access time. Technically, what this means is that one can put into an area of photographic emulsions that holds one ordinary image at least 20 (but perhaps many more) holograms of the black or white binary patterns needed for computer reading. Each of these patterns can be called up separately and almost instantly with programmed beams of light. One can record letters, words, ideograms, fingerprints, whole pages of information.

With the high-resolution photographic emulsions already in existence, one can record thousands of holograms on a small plate. As Professor Dennis Gabor points out, a single page of holographic microfilm could contain the entire *Encyclopedia Britannica,* and that, he asserts, is "not an exaggeration." Contrasted with other methods of information storage, this capacity is enormous.

Trying to cash in on this potential are a number of leading laboratories including Bell Labs, IBM, Xerox, RCA, and others. Probably the leader in this exploratory work is Bell which has had a number of groups working on it.

What they have been trying to develop is a peripheral memory or file system for computers that would hold in holographic form machine-readable information. Robert Collier at Bell points out that with the rapid random access of holography, it should be possible to read out whole pages of information in microseconds, which makes it a very attractive competitor to present disk and drum memories. (As of now, Bell engineers claim to have worked out the beam selection technology that allows 10,000 pages of information to be held in storage.) The method could also compete with present optical microimage storage because of its capacity to hold information despite pinholes, dirt, etc. In general, holography offers a good medium for any application requiring a recording medium that is relatively insensitive to noise.

So far, read-only memories have been developed, though there have been some interesting experiments with thermally erasable memories, with information stored in three-dimensional crystals, and the like. Of these experiments, the thermoplastic holographic memories (information is deposited or erased with heat) look most promising. It has been found that information can be recycled 100 times without degradation, although for the really long term it is not known what will happen as the plastic ages, or what kind of ultimate storage densities can realistically be expected.

Xerox has also done extensive original work in the area of thermoplastic memories. The company apparently is very much interested in the application of holographic storage for computer-based educational systems, for remedial reading and the like. Such automated "teacherless" education offers a potentially sizeable market.

Another variation of such erasable memories has just recently been announced by RCA. Their method uses a magnetically polarized material. The heat from the recording laser beams, when they are in phase, is sufficient to change the direction of polarization at a particular point; when the two beams are out of phase, their energies cancel each other out, and have no effect on the polarized recording sheet. Thus, one has a form of digital storage.

This general recording technique might be extended as a recording medium for infrared holograms (in the optical spectrum although invisible). If so, it could have a dramatic impact on machine tool applications of holography. For many measurements, optical wavelength holograms are too sensitive for practical use, but infrared wavelengths (using lasers in the 10 to 100 micron range) would bring the sensitivity down into a region compatible with machine tool work.

Another of the major contenders in the holographic memory business is IBM. Although the company is close-mouthed about its work, it is clear from the literature produced by IBM scientists that one of IBM's major interests is in holography's mass-storage capacity.

Moreover, the company was the first to make computer-generated holograms (the work of A. Lohmann), which could be used in applications requiring the construction of 3D curves.

Although the researchers in holographic memories insist that the work is in an early exploratory stage, it is clear that such memories might find application in telephone switching systems, in the storage of all kinds of records, in the storage of libraries, and so on. Holography would probably not find a place in computer central memories where the manipulation of data is of foremost importance. (In terms of cost, holographic peripheral memories could be more than competitive—it is estimated that storage will cost only about 1/100 cent per bit, which compares with one cent per bit for existing magnetic memories.)

Another exciting potential of holography is the development of associative memories, which has long been a dream with all breeds of computer-oriented people. Such memories might be developed by putting information into storage with both recording beams, so that they would be stored "in association." In such memories, it is possible to recall an entire sequence of information or part of a sequence when just a fragment is presented to it. This capability is similar, say, to a human's ability to recognize another person when he has just caught a glimpse of a profile or when he has detected the faintest inflection of a word in a telephone call.

Incidentally, this property that holograms have of recalling whole figures from the merest fragments has led to some speculation that the brain includes some form of holographic mechanism in its functioning. We take this aside here to suggest how immensely superior the brain is to the pattern-recognition devices in existence, and what an exciting field remains to be explored.

Among those people who have elected to work recently on associative holographic memories is none other than the original inventor of holography, Professor Dennis Gabor.

In the visual information processing field, the mechanization of relatively straightforward pattern-recognition tasks seems closest to realization. A number of companies, for example, are developing practical systems for automatic fingerprint reading.

But the application that some people believe will create the first big commercial breakthrough for holography, and that will pave the way for a wider consumer use of holography, is in the credit card verification system.

This development, which reportedly has the keen interest of one of the major oil companies, could answer a major problem. One oil company alone has issued through its gasoline distributors over 18 million credit cards. Of these, at any given time, tens of thousands are either lost, stolen, or determined to be bad credit risks. The company loses millions of dollars on these bad credit risks each year. Moreover, if it could be ascertained on an easy continual basis that credit cards were valid, gasoline distributors might become gigantic outlets for other types of consumer products (TV sets and the like).

A credit card verification system that is relatively cheap, reliable, and easy to use, might be built around the hologram and cheap mass-produced lasers. Such a system has been under development by a young man, Kent K.

Sutherlin, who had worked on holography at the Lockheed Company in California. He has pursued this credit card verification system at a small company called Baus Optics in New Orleans, with the funding of ICV, a company formed to market the credit verification system. Sutherlin has recently moved the development to San Jose, California.

The ICV system depends on the tremendous information storage capacity of a hologram, its redundancy or relative imperviousness to dirt or even holes, and its amenability to fast random access.

With this system, a single holographic slide might contain up to 9 million credit listings. (Two slides would be sufficient to contain all the listings for the company cited.) And with it, a gas station owner, for instance, could check conveniently and in a matter of seconds whether or not his customer's credit card was good. New master holograms would be mailed to the gas station owner every week as the credit listings were updated.

Considering the present high rate of bad credit cards, and the growing amount of cashless business conducted in the United States, a foolproof system of this kind could pay for itself many times over.

Allied to the research on computer and pattern-recognition uses of holography are imaging systems whose applications could reach into many fields.

At the Bell Labs, as at IBM, researchers have been developing methods of imaging three-dimensional information generated in computers. One conceivable application for such a system is in architecture, where an architect can feed in design data to a computer and then can see on a 3D display the building that results.

These three-dimensional images from the computer are actually composite holograms, which are constructed by techniques that are similar to those developed at the California Institute of Technology by Nicholas George for a totally different purpose.

George, who is known for his leading work in diffraction gratings, has been using composite techniques to construct 3D holographic displays from normal stereoscopic photographs.

In his laboratory, George describes, in a voice made husky by constrained excitement, what he views as one of the central unsolved problems of holography today. How, he asks, can you make a real image that is realistic? That is, if you illuminate a holographic plate with a laser, you can see the image in an authentic way, but only one person or so at one time can look at the correct angle into the plate. On the other hand, if you bring the reference beam from the opposite direction, a real image in space is formed, but the observer has to train his eyes very carefully to see it, or an artificial screen must act as a focus at a certain point of the real image. The problem is that it is not possible today to view the real image formation over a very wide angle. Although many people are working on this problem, it remains, George states, a long-range problem that is more easily stated than solved.

But the solution of it ranks with the problem that animated Dennis Gabor originally—namely, that holography would lead to greatly increasing the resolution of an electron microscope, which is also today still unsolved.

In his pursuit, George has worked on scaling and resolution of holograms, and he has developed techniques of complementing well-known stereo techniques with holography. For large objects or scenes, for instance, it is not possible to use monochromatic illumination. That is, no laser is powerful enough to illuminate an entire mountain range for instance. One solution to this developed by George is a kind of holographic stereogram, a way of making holograms of very large objects by essentially piecing them together from smaller picture elements.

One very interesting possible application of this work is in the field of aerial surveys. In such work, aircraft make series of photographs from an altitude of 10,000 feet. Actually, pairs of pictures are made, and these stereo pairs are run through fancy optical apparatus to show topography in 3D which can be used for road-building construction projects and the like. Photogrammetry (or measurements from such photograph pairs), as the field is called, is now a very commercial field. But the type of equipment needed for reading out stereo photos typically costs about a quarter of a million dollars.

At Cal Tech, George and his colleagues are working on a method of making holograms to replace the stereo pair system. Basically, they have come up with a simple system that uses the stereo pair photographs taken in the normal way with ordinary illumination. But then in the laboratory these photos are converted to mathematically transformed holograms, which are registered accurately on a long strip of film. When these holograms of the photos are illuminated with a laser, the viewer sees the 3D view reconstructed. George says that such a system, which could eliminate the expensive stereo pair equipment, is a simpler way of getting a topographic map from the hologram. As of now, the Cal Tech group have finished the work on converting the photographs into holograms. The problem that remains is to make automatic readout equipment for making topographic maps automatically from the holograms—and have it all cost less than a quarter million.

If this project is successful, and George and his colleagues feel that it will be, the entire field of photogrammetry will undergo a revolution in the type of equipment used for readout. The system has already been used to convert lunar photographs taken by NASA-LRC Lunar Orbiter V to 3D.

In the consumer field, Rand McNally has shown interest in using the Cal Tech techniques for making large visual displays of such features as the Grand Canyon.

Still another direction for holography is in the field of medicine, especially with a relatively new application called acoustical holography. This work uses acoustical waves to create the hologram, and then a laser is used to "translate" the acoustical picture into a visible one.

The method is likely to find some use in nondestructive industrial testing. Perkin-Elmer, among others, is studying acoustic holography for the detection of internal flaws and defects in castings, extrusions, machined parts, etc. But the more spectacular application is in the field of diagnostic medicine, where it is likely to lead to a whole new line of medical instrumentation.

For the fact is that acoustical holography could function as an x-ray without radiation hazards. Three-dimensional acoustical pictures of internal structures could be and have been made, as illustrated in the photos of the Santa Barbara work, under Glen Wade.

Work on acoustical holography has also been carried on quite extensively by Alexander F. Metherell at the McDonnell Douglas Advanced Research Laboratories in Huntington Beach, California. More work, by Byron Brenden and Gary Langlois, has been done at the Pacific Northwest Laboratory of the Battelle Memorial Institute. This work, incidentally, was sponsored by the Holotron Corporation (of Wilmington, Del.) mentioned earlier.

So far as is known, the first work in acoustical holography was done by F. Thurstone at Duke University.

Thus far, we have been talking about potential applications, developments that promise large markets, but which also still face large engineering problems. It is when we come to the industrial scene that we find some applications, not as large in terms of market, that are already in existence or very close to it.

One of the most promising and near-at-hand applications of holography, one that is being pursued at many places, is interferometry. Holographic interferometry is important to consider at some length because, of all the known applications, its discovery, its areas of application, and its unknowns, tell us most about the pregnant characteristics of this unique tool.

Holographic interferometry was discovered, first of all, nearly simultaneously in several different laboratories; and in each case by the same "accident," an accident labeled by Powell and Stetson (who first noted it in December 1964 at the University of Michigan Radar and Optics Lab and reported it in April 1965 at the Optical Society Meeting) as the "fortunate misfortune."

In their case, they were striving to produce really good quality holograms, which meant being almost "over-careful about everything." Just when they thought they were getting to be expert at this, and getting "some swagger," they turned out a hologram that showed a set of black streaks or marks. In disgust, they threw the hologram into the wastebasket, but retrieved it ten minutes later as it dawned on them that something else was up other than a bad portrait of the small metal train model they were holographing. But what? It took a couple of days of discussion as they drove together back and forth to Willow Run to set up a systematic search for what in their experimental apparatus, rather than in the resultant image, was causing the so-called black marks.

And they soon found it. To stabilize the metal train they used as a model, they had set it in melted wax so that it would adhere firmly to the metal slab base. But the metal base, which had been heated by the melted wax was still creeping when Powell and Stetson made their hologram, resulting in the black marks.

What they had stumbled on was an old principle exhibited in a new form. They had registered on their holographic plate an interference pattern set up by the microscopic shifts that had occurred in the subject during the exposure of their hologram. Very soon, they developed a technique for deliberately vibrating their subject while making the holograms and studying the kinds of interference fringes that resulted.

Almost simultaneously and independently, holographic interferometry was discovered at TRW Systems in Redondo Beach, California, by L. O. Heflinger, Ralph Wuerker, and their colleagues while they were taking pictures of bullets. They were pioneering a new technique for stopping motion in their holograms by using a short pulse ruby laser instead of the continuous wave gas laser. In their case, the pulse laser malfunctioned, fired a double pulse, and "out came an interferogram for free." After a few false starts, the TRW scientists put their finger on the cause, then went on to develop their double pulse technique.

In England at the National Physical Laboratory, J. Burch and a student were also astute enough to recognize that one of their accidents was an interferogram.

Subsequently, as the reports of these discoveries began to trickle through the scientific community, Powell reports, many experimenters told him privately that they had had similar "failures," but had not realized the cause.

But the real point of citing these cases is to show that interferometry is so natural to holography that it couldn't have been avoided. And those who immediately began to exploit it were pushing what seemed to be a limitation of holography and turning it into a beautiful and useful advantage.

All an interferogram needs is two exposures. An initial exposure is made, but not developed, and then the object under study is deformed slightly, or subjected to stress, change of temperature, allowed to grow, or whatever, and a second exposure is made. The difference (any difference in its shape) between the object at rest and under deformation shows up as a fringe pattern, and it is, in fact, a highly accurate measure of the slightest changes or anomalies. It is so sensitive a measure that if a book were the subject of a hologram, the change in the book paper caused by the moisture it might absorb in one minute between exposures would cause fringes to appear. With the use of the pulse laser, interferograms can be made of fractional microsecond (manosecond) events.

Actually, 2D interferometry is an old tool, but its use took great care, and the results were often imperfect. Holographic interferometry allows its users to apply it under conditions in which traditional interferometry would never have worked. Moreover, holographic interferometry offers the unique capacity of recording interferograms of diffusely reflecting 3D subjects of a large size.

At TRW, the mechanics group is using interferometry to study deflections in turbine blades, the materials group is using it to study the behavior of materials in high-stress and high-temperature environments, the aerodynamics group is using it in a ballistic wind tunnel to study flow fields around satellite re-entry shapes, and the interaction of high-speed events, the electrochemistry group is using it to study the diffusion of one liquid into another, and so on.

These uses of holographic interferometry for nondestructive testing are still only in the R&D laboratory.

But in Ann Arbor, Mich., GCO, Inc. has already pushed into the development of industrial testing equipment. Its tire monitoring equipment is a particularly clever application. If a tire is inflated, and a first hologram exposure is made of it, and then you wait for a short time and take a second exposure, you will get an interferogram of how much the tire has moved. Places where the tire layers are not bonded correctly will tend to creep more, and will show up as a kind of bull's eye ring surrounding the place of the defect. Apparently, the method will detect defects twenty ply deep. Although the method is still expensive, and perhaps will receive its first major usage in the aircraft industry, its application should eventually spread to the automobile industry as well.

In the few years since the discovery of holographic interferometry, a range of applications, methods, and equipments have been evolved. For instance, the ruby pulse laser method, which TRW people suspect will be the most important in the long run because its high speed presents fewer technical constraints, requires roughly a $20,000 equipment setup. But low-speed events can be studied with gas lasers, which are far cheaper. Studies of this type could be done virtually on a garage floor with a $200 gas laser, two orders of magnitude cheaper than the pulse laser setup.

So far, TRW Systems has not made any arrangement to develop its interferometric equipment for a commercial market, but they have evidently been approached by outsiders to set up an operation for bringing this development to the aircraft industry as a commercial product.

An industrial application that is actually a form of double-exposure holographic interferometry is something called contour mapping. For instance, instead of putting an object under stress mechanically, the wavelength of the laser is shifted or (what amounts to the same thing) the object under study is immersed in a different medium (e.g., a gas with a different index of refraction). What results is a picture of the object on which are superimposed contour lines like those drawn on topographical maps. Among other things, this kind of contour mapping provides a way of making very precise measurements of contour changes.

At the research laboratory of the Hughes Aircraft Company in Malibu, California, where Theodore Maiman first operated the ruby laser, several holographic projects with industrial potential have been underway. Needless to say, the Hughes scientists are interested in developing the interferometric testing aspects of holography. But of primary importance has been their work on a giant pulsed ruby laser system.

At Hughes, Maiman's invention had been followed immediately by the concept of the Q-spoiled or "giant pulse" laser developed by R. W. Hellwarth. With current versions of the giant pulsed ruby laser system, it is possible to get illumination over a greater area, although its more important aspect is that it provides both a higher coherence length and a greater power. The practical importance of coherence length is that it is analogous in a hologram to depth of field. The greater the coherence length, the greater the field that the hologram can record. Depending on the laser used, this field can range from 30 cm out to thousands of meters, as with the giant pulse laser.

This really relates to another characteristic of holography, which distinguishes it from photography, and that gives it a unique competitive advantage. Holography conquers the depth-of-focus problem!

In normal photography, if one is to study a complex event in space, one must decide beforehand what level of the event is to be in focus. Thus all surrounding events will be out of focus, and essentially one gets no record of them. But with holography, one hologram captures and freezes everything in a given space. Once the hologram is made, the investigator can study different slices of it at his leisure.

This aspect has already been applied in the study of rocket exhausts, in which the spatial relationships of the emerging high-speed particles can be examined in a way that no other technique allows. Thus, the method can be used for evaluating the performance of all kinds of nozzle devices—paint spray guns, oil burners, fuel injectors, and the like.

Another application, in which TRW has collaborated with Scripps Oceanographic Institute, is in the study of marine plankton. In this case, the holographic equipment is lowered into the undisturbed environment of these microscopic creatures, and a record is made of the entire population occupying a cubic meter of sea water. Once these holograms are made, oceanographers can begin to look for significant statistical patterns (e.g., does each plankton hold a "territory" he defends against neighbors, etc.).

Also at Hughes, still in an earlier stage, is research on new materials for recording holograms. Not many months ago, in fact, D. H. Close, Alex Jacobsen, and their colleagues, reported on a new photopolymer material that could be played back as a hologram after a fraction of a second. Moreover, this virtually real-time recording material does not require wet processing. If its further development is successful, it could open the way for many new holographic applications—on-line process control, display systems (for instance, a radar screen could be recorded every ten seconds or so and projected on a larger screen), and so on.

A third and more minor program at Hughes has been the use of their giant pulse laser to make a real 3D movie of fish swimming about in their tank. Although the movie was only a demonstration, it showed that the techniques used (a repetitively pulsed laser, etc.) do work. They point to the possible application of such movie systems to holomicroscopy, to time-revolved holographic interferometry and to time-resolved studies of particle dynamics.

Still another application of interferometry arises in quality control. Burch, in England's National Physical Laboratory, who was one of the early discoverers of the phenomenon (and who has since applied it to metrology), has used it in this fashion. By taking a hologram of the prototype piston or cylinder from an automobile (from a Rolls Royce, of course), it is possible to use it as a comparator against subsequent production models. The appearance of any fringes gives an accurate measure of how close the production model is to the perfect prototype.

The method is also being used in the United States for quality control and testing of all kinds (as in the testing of integrated circuit arrays which are essentially two-dimensional).

As well as for testing, holography might also be used in making the photomasters for the production of integrated circuits. With the present methods of contact printing, a speck of dust or scratches on the photo master can considerably reduce the yield of circuits from a single silicon chip. With the large integrated circuits (for memories, etc.), which contain up to a hundred major elements, the yield problem becomes very serious. But with a hologram mask, there is no need for contact between the master and the chip. Moreover, even if there are specks of dust or scratches on a hologram, the "entire message" gets through anyway, because all parts of a hologram carry the whole picture.

In that quasi-world between academe, industry, and the mass consumer market, are those many people who see into each camp and who see business opportunities at what they call the "interfaces." They do not want to see the magic of new technologies locked up in a few laboratories—they want to expose the general public to it, and they want to bring the impressive technology percolating slowly in the laboratories more quickly into the commercial world of usable products.

There must be many such men today jittering with excitement and frustration at the periphery of holography. The applications they envision extend from relatively simple toys all the way up to the most sophisticated devices of mass entertainment.

One of them who has seen an opening for doing his thing is David Bond, the president of a new small company in Torrance, California, called Holovision. Bond's first thing is a kind of mass-produced scientific hobby kit—a "holography teaching unit"—with which bright 5th and 6th graders, and upwards, can learn about coherent optics, light refraction, and holography. The kit, priced at less than $10 for a mass school market (anywhere from 50,000 to a million units) contains an instruction manual by A. V. Baez (yes, Joan's father), an early enthusiast of holography, and who is today an advisor to UNESCO on science education. Bond says he became convinced that holography needed more such exposure when he spoke at a science teachers administrators meeting and discovered that one half his listeners had not even heard the word "holography." Perhaps this shouldn't be surprising. George Beadle, the Nobel Laureate geneticist once said that the only audiences whose knowledge made him uncomfortable were either research scientists or teenagers (they asked the perceptive questions). (How many of our sophisticated readers heard of it before picking up this book?)

Another of Holovision's products is a viewing unit for looking at holograms, the whole unit being mounted in an attache case.

Of those people in the universities who took up holography immediately after Leith and Upatnieks showed their spectacular pictures were J. W. Goodman and M. Lehmann, and their colleagues at Stanford University on the San Francisco peninsula. These men have pursued a number of interesting directions, including the digital formation of images from electronically detected holograms (others like Brown and Lohmann and like the Bell Labs work mentioned earlier have used digital computers in the computation and construction of holograms). They have also worked recently on experiments on forming holographic images of very distant objects with a view toward eliminating the distortions of the intermediate atmosphere. Conceivably, such systems would find application in surveillance systems (of circling satellites) and in ground-based astronomy.

More intriguing are some of Goodman's and Lehmann's extracurricular holographic activities. (They are, by the way, consultants on the credit-card verification system discussed earlier.) One of the projects that Lehmann has gotten involved in is with a Santa Barbara company called Creative Imagery, which is trying to perfect a technique for making holograms of art works, so that traveling displays of art works (sculptures, etc.) could be sent around the world without the originals ever leaving their museum homes. Lehmann has gone so far as to design a holographic laboratory mounted in a trailer truck. The truck would simply be backed up to the museum loading platform so that the art objects could be holographed right on the spot. With the perfection of such techniques, it is possible to foresee a day when many sculpture exhibits could be sent traveling virtually in an airmail envelope.

Lehmann has also been approached by a New York advertising firm which is interested in working out similar techniques for jewelry displays. Lehmann, a modest, bright-eyed man, chuckles as he imagines the consternation of a jewel thief, as in the story, "Topkapi," as he reaches in to clutch the prize and his hand closes on nothing. With these two examples, our imaginative readers will dream up many other such possible applications.

Last but hardly least in the list of possible applications for holography are those that one might have expected to have been developed first—three-dimensional television and three-dimensional movies. For these, surely, there must be a gigantic market.

But actually these applications probably present the most difficult technical and the most costly economic problems to overcome.

RCA has announced a method of producing prerecorded television programs on tape that involves the use of holography, but even this use has many technical problems to overcome. The company is gambling many millions of dollars on the venture (which seems like a response to the EVR system of CBS), which certainly should give the holography art a boost, but it is not aimed at making 3D television.

About that prospect, in fact, Dennis Gabor is pessimistic. He once tested a 3D projection system on the scale of a television screen, but he found one could not take the images seriously. "It looks like a puppet theatre," he says.

But Gabor has continued to work on many aspects of holography, such as three-dimensional art works, and on a technique for a three-dimensional movie. And this he does take seriously.

This latter might even be a movie that completely surrounds the viewers. Gabor has a patent pending on it—"it is the perfect solution for the 3D movie problem"—but the necessary funding to develop it, he admits, is so large that the project might need to wait until the 21st century!

Meanwhile, the inventor of perhaps one of the most intriguing inventions of this century must continue to wait to see the commercial outcomes of his brainchild. His rewards thus far? His fame, his election as a Fellow of the Royal Society, a modern "three-dimensional villa in Italy" where he spends two months each year, and the satisfaction of seeing hundreds of others now mining out the unknown ores from his already prolific single invention.

Perhaps, too, he will yet see someone step forward to put up the X millions of dollars needed to create the 3D movie before the 21st century.

November 1969

Part 2

Some Strategies and Observations

9

Bringing Technology into Biomedicine

In many technological fields, it is the marketplace that drives the innovation process. But this is less true in the emerging technology of biomedicine. In this chapter, we look at the process through which basic biomedical research gets translated into acceptable instruments, devices and systems for health care.

James F. Dickson III analyzes the maze of obstacles that holds the field back. If we are prescient enough in our present planning, he says, if we can manage to manage the dynamics of the process, we might, just might, *stagger our way into that Brave New World, the medicine of tomorrow.*

Mr. Dickson is a Program Director at the National Institute of Health.

People in the United States are not getting as much benefit from the new knowledge of medicine—the new technology of health—as they could get or should get. Inadequate planning in the past, inadequate funding in the present, resistance to change, rising population in a period of infinite expectations and limited resources, awareness of the powers of technology, the steadily rising cost of maintaining and protecting one's personal health—these are all factors increasing the pressures to improve greatly all aspects of the system whereby people can get adequate medical care at a cost they can afford.

Like other fields today, medicine, hard pressed, is reaching out for help from engineering and from industry. And there is much that engineering is doing and can do to ease the pressures on the present medical system. But the problems of the biomedical field are peculiar ones, and engineering cannot help merely by wishing to, nor can industry make easy profits simply because there is a rising demand, for the market for medical devices and systems is unique unto itself. New ideas—innovations if you will—must fight their way to usefulness through more than the usual maze of obstacles, and tremendous energies are consumed in accomplishing damned little.

What we are dealing with in medical and engineering developments related to medicine is a highly diverse governmental-industrial-academic complex. To get anything done is a matter of politics and sociology as much as science and engineering.

It is enough of a problem just to decide on our goals and priorities, but to get it done is the real problem. It's not going to be easy, because the dynamics of the process of translating ideas to results isn't well understood. Our knowledge and insights into how engineering can be brought to bear effectively in all aspects of biomedicine is still at an early stage. To draw general rules for the management of this process is very difficult and perhaps premature.

The plain fact, so far as I can see, is that very few people today have much worthwhile insight into the ways in which basic biomedical research gets translated through development into more acceptable instruments, devices, and systems for health care.

What I propose to discuss here is what we do know about the process of technological innovation in the biomedical field, some of the factors that hinder it, and some of the ways that more satisfactory activities in this field might be stimulated.

This means that I need to talk about how engineering relates to the world of biomedicine in a very general way, how both have been changing, the need for better planning if we are going to get the medicine of tomorrow by tomorrow instead of by 1990, the consumer industry and how to stimulate the market, financing, manpower, education, national laboratories, the role of the federal government, patents, health as a political problem, how we set priorities, our pluralistic society, and even the character of different types of doctors, to say nothing about the poor guy at the end of the line, the one who is looking for a physician who can help him get well or stay well. In essence, I want to talk about the evolution of the technology that can free the doctor and return him to the patient. This is a tall order for a short essay, so let's get going.

Although I wish to speak in fairly general terms, I should like first to give the reader a feeling for the relative magnitudes of the pieces of the industry we are examining.

At the present time (in fiscal 1971), the total health industry in the United States—the hospitals, physicians, producers of medical supplies and equipment, everything—amounts to about $67 billion a year, the third largest industry in the country. Estimates are that by the middle of the 1970s, it will be a $100 billion industry and the Number One employer of people in this country.

By contrast, the biomedical engineering industry (or the medical electronics industry as it is popularly called), with which I am most concerned here, amounts to about $350 million, or less than one-half of 1% of the total medical industry. One must realize that these figures are very rough because there is no central compilation of statistics in the medical industry.

However, it is estimated that this medical electronics industry should grow at the rate of anywhere from 5 to 20% per year, depending on many variables—the level of government funding, the course of the Vietnam War, the fight against inflation, etc. During the coming decade, biomedical engineering should constitute a $1 billion per year business.

Within this business, one can point to three distinct markets: 1) biomedical research supported mostly by the federal government, largely through the National Institutes of Health (NIH); 2) the teaching market, which is relatively small, consisting of devices for training and the like, supported partially by private philanthropy, industry, and government; 3) the clinical market, which has the greatest untapped potential and which, in contrast to research, receives much less federal money support—roughly a third of the clinical market is supported federally, a third is supported by hospitals (borrowing, depreciation), and a third by foundations and private philanthropy.

The real consumers in the growing clinical market are the hospitals. There are 7,000 hospitals in the United States, all potential buyers of the new clinical equipment. Five years ago, only the largest hospitals were buying such equipment, but now the smaller hospitals are coming into the market picture.

In the Department of Health, Education and Welfare, in recent years there has been some increasing emphasis on developmental work. For instance, under NIH, which is part of HEW, artificial hearts, artificial kidneys, automated clinical laboratories, and assorted instruments have all been under development. HEW has also set up the National Center for Health Services Research and Development, which has gotten under way in the last couple of years, and a part of whose mission is to develop a variety of instrumentation for health care. This Center will be an important focus for all such work.

The range of NIH's programs in biomedical engineering research includes such things as bioinstrumentation for the prevention of disease, diagnosis, treatment, and rehabilitation of patients; the application of engineering control concepts for the study of biological systems; biomaterials; engineering research pertinent to surgery, anesthesiology, diagnostic radiology, and the care of the critically ill; computers for medical diagnosis, patient monitoring, and hospital information systems; orthotic/prosthetic devices; and the automation of clinical laboratories.

A rough estimate of the total funds that NIH puts into biomedical engineering (including R&D and training) is some $50 to $60 million per year.

What is happening to us these days, as everyone knows, is that the population is increasing, there are rising expectations on all sides, and it is clear that we need more satisfactory techniques in health, across the board, for the prevention, diagnosis, treatment of disease, and for the rehabilitation of those who have been afflicted and treated.

At the same time that these needs are intensifying, health costs are rapidly mounting, and they are accentuated by general manpower shortages in the health field. Those in the medical field are aware of pressure from many sides to bring about more efficient application of present knowledge and technologies in a way that is tangible, in a way that people can see. People now recognize that investment in biomedical research for the past fifteen years has brought about new knowledge, new techniques, new technologies, but they see very little of this "new" leaking into neighborhood medical practice, especially in our cities. So this combination of two things—the people's increasing awareness and the skyrocketing costs of hospitalization—is forcing the question of health services to move more to the forefront as priorities are established by the Executive Branch of the federal government.

In the midst of this, the role of engineering in the world of biomedicine is becoming more apparent. As we all know, the character of engineering has been changing over the years. There has been an influx of science into engineering education; and one result of this is that engineering has become more welcome in the basic and applied parts of biomedicine.

It used to be that engineers could make utilitarian instruments to help doctors, but the newer engineering, equipped with theories of communication and control such as those developed by Norbert Wiener, has become important to basic biological studies like the control of bodily homeostatic mechanisms.

As one looks at the problems of health care in the coming ten years, one sees that our engineering capabilities can be directed into two major areas. At one end of the spectrum, it can aid in obtaining knowledge about life itself, in basic biomedical research; and at the other end, it can come to grips with the everyday health needs of all the people.

Fifteen years ago there was virtually no engineering in medicine, and in fact there were no formal biomedical engineering training programs. But,

relatively speaking, there was a lot of seed money to get things started. Now there has been a considerable impact made by engineering throughout medicine and there are some eighteen biomedical engineering doctoral training programs in the country. Further, instrumentation has been developed on a broad scale and is ready to explode into several clinical areas.

However, the money situation at this time is relatively discouraging in spite of biomedical engineering's ripe potential. What is clear is that if biomedical engineering is to be accented in any major sense, additional funds must be provided for it.

It's likely, owing to the rising demand for better medical services, when people regard proper medical care "as a right and not a privilege," that the application of engineering to biomedicine will grow greatly in importance, and that this whole area will get moving again in a few years.

But the hazard is, in the current hiatus in funding, that the pressure may grow on the biomedical engineering fraternity to show some results, any results, of biomedical research, so that people will be more inclined to let money flow into basic research. However, it does seem to me, from where I sit, that it would be a mistake to rush out and attempt to do all the things one might do if there were adequate funds. In such an atmosphere, there could be sloppiness and a poor selection of projects.

An even greater danger is that, under these pressures, funds could be funneled away from basic biomedical research to pay for engineering developments just to show that there are tangible results. I believe this kind of robbing Peter to pay Paul would be a basic mistake.

Nevertheless, I see the current hiatus as fortunate in one sense. What we have is a forced time for regrouping our thoughts, for reflection about what we are doing in biomedical engineering. It gives us time to formulate our understanding about the whole interactive field and to arrange for better planned programs.

I really believe what I've said is true. I'm not merely saying "When money gets tight, decisions get better," as they do in the home or in any organization. That's true in part, but not what I'm talking about at all.

What I mean to say is that, because we don't understand the dynamics of the process of translating basic biomedical research into effective health systems, we can't surge ahead effectively anyway without the most careful planning. So we now have a period when biomedical engineering can engage industry in a gentle way, to find out how it can make its most effective approach into the development of biomedical devices and systems. For the fact is that, despite its increasing productivity of medical items, industry's efforts in biomedical research have not flourished.

I'll go into some of the subtle reasons for this and have suggestions to offer later in this essay.

Meanwhile, we who are interested in this field, inside the government or out, people in technical management, would be wise to spend our time improving our understanding of what is really happening in the biomedical field, to make better plans and to argue out what is obscure in open court. I think of this essay as a contributing salvo in this dialogue.

Because there are different ways of looking at the world of biomedicine, to make some things easier to conceptualize about its relation to engineering, I would like to refer to the drawing here that shows biomedicine broken down into three major areas—research, development, and delivery of health services.

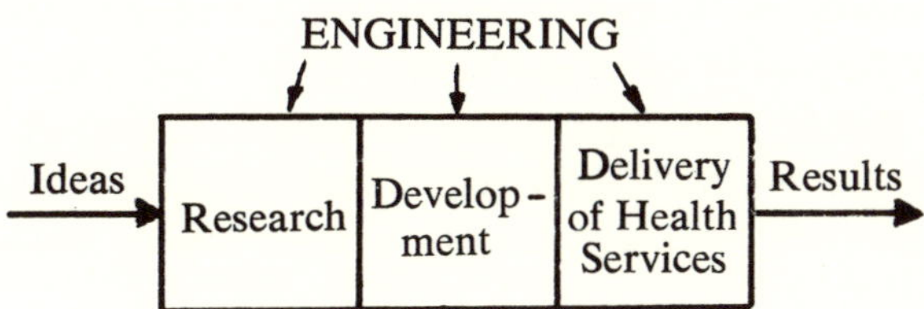

The top of the diagram shows that engineering has something to do with each of the major biomedical areas. One of the possible ways of looking at biomedicine is to see it as a throughput system—someone's basic idea is carried through a maze of research and development and finally transformed into something that relieves people's pain and distress.

But this idea-to-result is a dynamic ongoing process, with lots of feedback loops between all stages of the process, with time lags, and with both stimulating and inhibiting factors. What we are looking for are the ways we can properly stimulate this process, which is presently in a sluggish mode, to get it into what an engineer would call a regenerative mode. For instance, how can we tinker with biomedical research and development so that consumer manufacturers will be induced to come into the field both for their own profit and for the overall benefit of our health system?

Now the interesting thing about this picture is that engineering can help at every stage of this process of getting equipment out to where it is useful and where it can be used in a rational way.

In research, for instance, engineering concepts of communication and control are being applied to biological systems, such as the cardiovascular and neuromuscular systems.

A recent striking achievement in development is the work of Norman Anderson at the Oak Ridge National Laboratory in Tennessee, where he has been using concepts of centrifugation to revolutionize the approach to analytical biochemical studies. The instrumentation that he has developed has been gotten out to industry (in a program I'll discuss later in a more relevant context) and very shortly will be on the market. This new instrumentation will contribute directly to the automation of clinical laboratories and thus satisfy one of medicine's most pressing immediate needs.

And in the delivery of health services, one can point immediately to the elaboration of information systems and to systems for monitoring critically ill patients.

In the above sense, biomedical engineering can be defined as the application of the principles and practices of engineering science to biomedical research, development, and the delivery of health services.

Now that we have cleared away an overview of the relation of engineering with biomedicine, we can approach the heart of our present problems, which lie in the interaction of academia, industry, and government.

Let me start by pointing out a few features of university structure, industrial aims and the medical market, to demonstrate how their interactions and loop feedbacks have tended to repress the onward movement of biomedical engineering.

First of all, the prevailing university structure makes it difficult to take advantage of the opportunities that can be gained by interfacing engineering with biomedicine. For perfectly good historical reasons, universities are organized along departmental lines, whereas biomedical engineering is an interdisciplinary activity, requiring the cooperating talents of many kinds of specialists.

So this puts us into a kind of bind, because most of the best research talent is in the universities, but the academic structures which are outmoded with respect to interdisciplinary undertakings inhibit their being properly matched to the biomedical problems to be solved. Although a few universities have recently set up foci of biomedical engineering activity (usually as part of their engineering departments), we must recognize that university leadership by and large hasn't rushed to meet the needs in this area.

One might point out that the problem of outmoded departmental structures vis-à-vis interdisciplinary work is by no means confined to the universities. Most organizations in our society have been outdistanced by the forward thrust of technology, which more and more is acquiring an interdisciplinary character. This problem goes all the way up to the highest levels of government.

One bright side to the university picture, however, is that there very definitely exists a movement among the younger faculty members in the medical schools and engineering departments to develop more interdisciplinary courses with the bright younger students who are really driving the system through their interests and their questions.

A new school is being organized by Harvard and MIT that will be called the School of Medical Science and Technology. It will be all-encompassing, including the relation of mathematical and physical sciences as well as engineering to biomedicine. Most of the bioengineering programs up till now have been, in a sense, pilot efforts towards this next step, a generation of larger interdisciplinary schools of health, science, and technology. I believe this is an extremely important step forward.

Another aspect of university involvement in this field is that the type of manpower and resources needed to carry an idea successfully through prototype development is generally not found in the university. To do things right, the university would need the kind of structures, manpower, and resources that industry possesses in good measure.

On top of this, some university intellectuals tend to regard work on devices and instruments as pedestrian. To work on something mundane that might relieve somebody's toothache is not highly regarded and is not likely to get the investigator far in the hierarchy. So people who are thinking of their careers are reluctant to engage in such dirty fingernail work. At the other end of the spectrum, some of the systems jobs that need to be done are simply too big and too complex for the university to handle with ease. The result is, of course, that in such a setting, biomedical developmental work does not flourish. Of course, the universities could counter this by putting more of an accent on the development and delivery problems, but thus far the major emphasis has been on research.

This situation, then, derives you in the end to think of making amalgamations of university and industry to get things done. This means you must weigh the strengths and weaknesses of each, and consider on what kind of basis they would collaborate. What is each scared of? What can each do best? But to begin to amalgamate you need some basic information.

When you consider what industry has done or can do in this field, you find it also faces deep problems. In general, it hasn't been terribly effective. I don't mean that there isn't roughly $6 billion of stuff coming out of industry that relates to health, but, in terms of medical engineering development, industry hasn't been very innovative.

One of the major reasons, whether anybody likes it or not, is that industry doesn't usually possess the biomedical knowledge and creative skills it needs to move ahead.

So rather than being in the heart of the process of helping to push the field ahead, they tend to behave merely as suppliers.

What further frustrates industry involvement is the character of the medical market. There are three or four telling things about the biomedical market that make it very difficult to get into and different from other product fields.

For one thing, it is most often a high-need, low-volume market. The country needs 100 of something, and needs 100 of them badly. Who the hell wants his company to get into that?

Moreover, the market is generally very uncertain, hard to predict. Doctors are individuals, and act that way—one wants one thing, and another wants another. Marketing researchers can't really crystallize the shape of the future market. They certainly expect that there will be a market, and they recognize that somehow it's got to be an $x billion industry, but exactly what will the pieces be like? Nobody really knows that.

That being the case, industry is certainly reluctant about committing resources, especially at a time when engineering talents are in short supply; moreover, industry has other alternatives that are crystallizing and that are well identified.

The result is that the role industry has played thus far, in finding creative and innovative solutions to root problems of health care, is virtually zero.

So the consumer industry is looking for direction, wondering where the money in biomedical engineering is coming from, what are the goals and priorities, and who is stimulating the market. We've got to stimulate the biomedical engineering market. For instance, nobody would have gone to the moon if someone had not created the market. John Kennedy said, "We're going!" and then there was a market.

The market problem also hinges on another factor. Consumer-oriented industry is highly organized, and much of its success depends on the smooth distribution systems it has developed and the people who have been particularly trained to work with it. But in the health field the same distribution systems don't exist. On the one hand the people don't have the right access, and on the other must sell high-cost biomedical equipment. They have trouble supplying needed services, which can be crucial.

Studies have shown that from the first time a new device is used in a hospital until it really begins to move, marketwise, may be five or six years. This is due in part to the fact that industry is unable to control consumer acceptance in the medical field as it can in other fields.

This unpredictability and unmanageability of the medical market comes straight out of the nature of medicine, which is an art, and out of the character of doctors and medical practice.

Doctors need as much as ever, and perhaps more than ever, a broad awareness of human values and sound personal judgments. This is true. This is what taking care of patients is really about. This is the art we need more of, and hopefully it is what we will get when we bring in more instrumentation and automation. We want to free the doctor and return him to the patient. The implications of this are something that any industry going after the biomedical market should understand.

Which is not the end of industry's problems by a long shot. There are two other factors I want to spell out which have multiple consequences in terms of the feedback loops of the overall process.

Long-term financing is another major trouble for industry. The current financing cycle in the consumer industry, as I understand it, is on the order of eight to ten months, with the money turning over three or four times a year. Now this isn't realizable in the medical field because products cannot be gotten out fast enough to turn the money at this rate. So you have to find much longer financial commitments than the consumer-oriented industry needs, and this is of particular concern to the larger companies.

Despite capital's interest in the biomedical field, and their realization that it will be a big growth area, there are major problems to be overcome with respect to risk capital pools. However, I do know that these problems are finally beginning to be attacked by venture capital interests.

The last factor in this catalogue of industry's woes is manpower. One can look at manpower on three levels—technician, undergraduate, and graduate training—all relating to biomedical engineering, development, and production.

First, while technical training schools and junior colleges would like to train technicians for biomedical engineering, they are reluctant. Despite the demand in the health field, there is very distinctly a lack of job opportunities. For one thing, the salary level for such electronic technicians is about a third higher than the level for nurses, and hospitals are unwilling to pay them more than they pay nurses. Once the technical training schools got wind of this, their incentives to get into the biomedical engineering service were considerably dampened.

As I mentioned earlier, some undergraduate training programs in biomedical engineering have been established. Many of the students in these are in cooperative programs, spending alternately a number of months in an academic school program and a number of months in industry. But it is extremely difficult to place these students. Hospitals aren't organized to take them and make good use of them, and the service organizations of industry won't take them because they don't know enough about the field yet. Moreover, those who do get placed bring back the word to other undergraduates that the consumer-oriented industry is unwilling to push ahead vigorously into biomedical engineering. So these undergraduates are unwilling to engage in graduate work. No jobs. And those few who stick on through graduate work, and who ordinarily (in other branches of engineering) take postgraduate training in industry, also get stymied by industry's lack of involvement in the field. Those who do get through, therefore, tend to stay in the university environment.

Now one can easily see how this chain of reactions works. Industry gets depressed, education gets depressed, and the whole output gets depressed. What one sees is a multiple feedback loop with a distinctly negative effect.

I could actually draw all these loops on the diagram of how engineering interacts with the world of biomedicine, and it would look horrible, but it would depict the actualities of the process. The market is simply not driving the system as it does in other technological fields.

And the need, as I see it, is to find ways of creating more identifiable markets for industry, and to provide industry with long-range reliable commitment of funds. The stimulation of clearly identifiable markets, and the establishment of priorities, I see as a function of government. Which puts another big loop in the process.

One could conclude from the foregoing that the federal government should help determine product requirements and specifications. Which is the same thing as saying that there should be standard designs that companies can manufacture into. Then all the loops in the system, ideally would start feeding the other way. At least in theory.

Government could also arrange contracts for medical engineering developments so that development and manufacturing could take place on a long-term financial commitment. Such financial commitments could be enmeshed with other risk capital pools, state money, and so on. There is not just one answer.

This tack, however, immediately raises the question of patent rights. Who would get the benefit of inventions made under such government sponsorship? In answer to this question, I'd like to make one sweeping statement here, and then elaborate on it later. In general, the patent situation, from the point of view of industry's incentive, is rather satisfactory in light of the manner in which the Department of HEW is actually administering it. There are complaints; it could be improved; but it isn't really bad.

What presents more of a problem is the setting up of some form of certification of safety for devices, instruments and systems. This is especially important in terms of total packages. Many of the accidents that do occur are not due to the manufacturer's package, but to the fact that it was hooked to equipment of other manufacturers for which it was never intended. The manufacturer may be innocent of negligence, the doctor who uses all these equipments may be innocent, but the accidents occur, because no one has set standards for safety, reliability, etc.

Naturally, then, manufacturers are scared of being sued. Studies show that this is really one of the problems of the consumer industry in getting into this field. So there must be certifications with respect to biomedical equipment.

One solution might be to have certification done through a nonpartisan laboratory. Electronic manufacturers, for instance, maintain underwriter laboratories. But there is no parallel for this in the medical field. However, it has been suggested that the National Bureau of Standards or the Oak Ridge National Laboratory might assume some such responsibility.

However, at this time, the whole business of certification is tied into the larger problem of regulation and control of all medical devices and instruments. This question is being considered by industry and medical societies and at legislative levels in government. The pressures are on, and we may be seeing some new form of overall legislation in the next few years.

Another way of helping to resolve this question comes back to the market. The fact is that a major portion of medical equipment in the next ten years is going to be purchased, either directly or indirectly, by government funds. So manufacturers themselves have suggested that such funds be applied only for equipment that is certified.

Thus far, I've talked only about federal money. But there should be some form of local subsidy for health projects that are not profitable per se. This would mean that some states would require changes in their banking laws to allow banks to form risk capital pools, which then could put up money for the long term to support those organizations that are willing to run the risk of the high-need, low-volume market. Such legislative changes one sees are another loop in the process.

An example of this kind of process is an organization that was formed about two years ago in Illinois. Called the Illinois Biomedical Engineering and Research Corporation, sponsored by the Governor of Illinois' Science Advisory Council as a nonprofit organization, it has offered to act as a clearinghouse for money that might be aggregated to finance prototype development in that regional area. (It has also proposed legislative changes of the type mentioned in the preceding paragraph.) Although it's an intermediate group, it could serve many functions, and become a sort of way station for biomedical innovations.

This organization in Illinois could itself be a prototype of a quasi-official way of tying into a more central governmental body. For instance, some manufacturers have suggested the establishment of a National Health Council in the Executive Office of the President. Such a council would oversee the health-related activities in the different agencies of the government, including HEW, and could through its committees help select prototypes of new medical equipment that should be built. This council would sit high enough to see all the troubles nationally, and would commit the work on such prototypes to different national regions that possess the appropriate resources. Groups like that in Illinois would conceivably arrange for further developments, testing, and production.

Another interim strategy for getting biomedical engineering moving in this country, while manufacturers are waiting for the patent situation to improve, is being tried. It amounts to separating preproduction prototype development and testing from actual industrial production, and, in effect, frees industry from the expense of prototype development, without unduly hampering industry's entrepreneurial spirit. The process is simple: Prototypes are developed in a nonindustrial laboratory which is well integrated with basic and applied biomedical research; then when the prototype is carried far enough along to give assurance that it is workable and that there is a profitable market, industrial people are asked to bid on it for manufacture.

The National Institute of General Medical Sciences has employed this approach for the development of automated clinical laboratory equipment undertaken by Norman Anderson's group at Oak Ridge National Laboratory. The prototype was developed at Oak Ridge and tested in the university environment. The government holds the basic patents.

The next step was to call a meeting of industrial people where the equipment was demonstrated. Industry's option was to take the basic prototype, embellish it as they pleased, hook it up with computers, or whatever their special interest might be, and then manufacture the equipment and sell it competitively. The companies were free to patent any improvements or embellishments of the basic equipment. As a result, four or five industrial groups picked it up—Union Carbide, Electro-Nucleonics, and others—and are now turning out different versions of this fast analyzer equipment.

Here in microcosm was done what might be done on a national scale for many types of biomedical equipment. It constitutes a model of how industry can get into the biomedical picture.

I must point out that each type of equipment presents its own special problems. In each case one has to sit with it, baby it, hand-tailor it. And each one is a different adventure—the politics, the money, the market, the underlying science are different in each case. This is the fundamental characteristic of this type of work. It's not like incorporating a new technical development in an assembly line of television sets.

For one thing, one must be very careful about how to proceed. A lot of people have wild ideas, but the scientific base for what they have in mind may not be adequate In the case of the automated laboratory equipment, for instance, we knew that at the bottom it was based on solid, new, and exciting analytical chemistry techniques. And now the manufacturers know they have something that is pretty damned good.

The important thing is that this is the way to stimulate industrial participation. If the manufacturer had to run the risk, the whole thing would slow down.

This mechanism of using preproduction laboratories and then farming out production is, of course, not going to work in every case. What this means is that the engineer or the industrial manager who is going to reach into this field must get some feeling for each individual situation, each type of doctor, and so on. He must become something of a sociologist. He must know something about how medical laboratories are run, learn something about hospital politics, know that selling automated clinical laboratory equipment is different from selling an electrocardiographic machine or a stethoscope to a general practitioner.

You've got to know doctors. Those in hospital laboratories are caught up in problems about money and about control. Surgeons are something else again. In college, the surgeon was probably a football halfback; he likes to use his hands. He's like the engineer, which is why they get along so well. They're both doers. In general, they get the satisfaction by jumping in, cutting and solving, often acting fast to save the patient. Internal medicine men are intellectuals, pipe smokers, puffing around and thinking it all over.

Now these descriptions may sound like truisms or exaggerations, which they are, but you'd be surprised how few engineers have any acquaintance with such gross features of the medical environment they are working into, much less the really subtle features.

Another fact is that doctors know very little about engineering, and engineers know equally little about bio-value judgments in medicine. This gap in mutual understanding has been lessened in certain highly select circles, but in general it persists and presents a long-range educational problem. After ten years, one gets sick of always mentioning this problem, but it still exists.

Of course, there are those systems type thinkers who will argue about this business of the multiple medical market. They point to the development of some type of biomedical instrumentation that will somehow go into every house, like the telephone, and give the whole biomedical industry a big shot in the arm, bring a lot of money in, and drag a lot of minor works along with it, that is, the speciality items. This may be true all right, but these people are talking about 1990. Whenever I get the chance, I keep the hell out of anything beyond 1975. To talk about the great blue beyond when I can barely get home at night just gives me the shivers. I prefer working at something I believe is feasible now within the dynamics of the technological process of innovation as I understand it.

What I try to keep continually before my eyes is how we can help preserve and increase the flow of support for basic and applied biomedical research by expediting the flow of results out of such research . . . so that it relieves people's pain and distress in the here and the now.

Just how fast is biomedical research and development likely to move ahead? In general, I believe, the activity in the applied R&D phase over the next several years is going to expand nationally at a very conservative pace, for the very reasons I spelled out earlier . . . including a shortage of money, locally and nationally.

Even though many people now want to use their engineering in the biomedical field, the fact is that only here and there is the scientific base really adequate to the risk. In effect, I am saying that the opportunities are really less than they seem. Those who wish to undertake serious projects will do well to measure the depth and breadth of the scientific base before they get involved.

A third variable that keeps a brake on the system, even if money is forthcoming and a decision is made to do something, are the people who actually do the development work. To get an integrated, effective group of people who possess the particular talents needed on a biomedical design and development project is something that doesn't happen overnight. When I spoke earlier about preproduction work in a federal laboratory, I may have sounded as though I were saying, "We just go ahead that way and do it." It isn't that easy.

The organization and the character of the people for an interdisciplinary project is a key issue. Typically, it takes several years to build up the right group that can handle both engineering and bio-value design judgments. That old bugaboo about the lack of communication between different types of specialists is always lurking around the corner. So this is another lag in the system.

Moreover, you need to know what is required of a group to be successful in this field, so that you can recognize them when you've got them. First, the leadership or management has to be interdisciplinary in itself. If the guy at the top isn't both a good physical and life scientist, the guys at the bottom will be laughing at him because he'll be making stupid decisions in one or the other specialty. Yet such interdisciplinary leaders are still rare animals. Another time lag. Second, the group needs solid financial support to ride the severe ups and downs typical of this work. Third, the group needs a good systems engineer to put all the diverse pieces together in the right way.

Because there is such a shortage of resources for the biomedical engineering enterprise, there has been some interest lately in reconsidering the role and resources of our national laboratories. There are scores of such laboratories in existence, with many capable people working in them. The activities of these laboratories are generally arranged around the particular missions of their parent agencies. But over a period of time, many of these labs have outrun their missions, and are looking for new territory to work in.

The question is whether or not such capabilities can more generally be appropriately matched up to biomedical problems, as in the instance of centrifuge technology at Oak Ridge—a capability that, in fact, existed nowhere else.

These laboratories are actually a national resource, and it is being asked how one can best use these resources in the national interest.

Throughout this entire chapter, I have been talking about planning—how to plan for a higher rate of innovation in biomedical technology and therefore deliver better health care. In discussions about planning on a national level, relating to problems that cut across both the public and private sectors, I have been asked how our overall efforts compare with those, say, of the Soviet Union in which planning is central to the system.

If one surveys the instrumentation available in the United States today, one sees, with some exceptions, that it consists largely of transistorized versions of instruments that were available ten to fifteen years ago. This situation exists in spite of the fact that many new techniques have been evolved that hold promise for a major upgrading of the methods associated with patient diagnosis and treatment.

A similar situation existed within the Soviet Union prior to 1962. However, at that time, the Soviets established a number of laboratories whose primary role is to screen instrumentation techniques developed in ongoing basic biomedical research and to identify those that hold promise of benefit to the community as a whole. Once instruments are singled out, the laboratories construct a limited number embodying these advanced concepts in such designs as can be used by relatively unskilled individuals. Follow-up on the performance of these prototypes leads to further refinements. Once the utility of an instrument is established, the design is transferred to manufacturing facilities for low-cost production and widespread dissemination.

Although it is difficult to get accurate information about the instruments being developed (one hears about breakdowns, difficulty of maintenance) under this approach, one can see certain generic advantages and disadvantages. Because biomedical engineering in the U.S.S.R. is driven from the top, the man at the bottom has little incentive; whereas in our pluralistic system, it is difficult to drive the system very well from the top, but the man at the bottom who wants the action can drive it.

I could go into this comparison in great detail, but I will leave it with one conclusion. On balance, I believe our system runs a little better.

With respect to the setting of priorities, we certainly do need more planning and more coordination of our limited resources, but we must respect the kinds of differences that I have been suggesting actually typify the biomedical environment, of both the doctors and their patients. So, although it is logical and necessary to start planning for technological advances at the federal level, I believe it would be a mistake to have just one group decide for us what kinds of medical instruments and systems we shall have. This may sound somewhat paradoxical, but I believe we need that leeway in the development of different kinds of medical equipment.

I believe we need to allow those people who are actually doing the bio-medical engineering development (those in the middle of the diagram of the biomedical world) to continue to be a pure reflection of the inputs from both ends—to reflect the constantly changing findings of bio-research and to reflect the responses of the individual patients whose health cause is being served.

You see, in looking at the process whereby we push for needed innovations in our biomedical technology, I am trying to argue for a strategy that is compatible with our pluralistic, democratic system. One reason is to get the job done within the context of the job itself. The other reason is to keep our system itself viable.

March 1970

10

What Really Happens When Automation Arrives

The use of computers to perform a large share of the labor in engineering design and manufacturing is far more than a dramatic advance in technology. It heralds a rearrangement of the entire company into one intimate process and, as a result, calls for some hard thinking about traditional ways of running industrial organizations.

O. Dale Smith lived through the forerunner of this revolution when numerical control entered the factory. Here he relates how the lessons that were learned then apply now. Only more so.

Mr. Smith is on the technical staff of North American Rockwell.

Year by year, the digital computer keeps bringing other revolutionary processes behind it, each one capable of shaking an industry. One example is punched tape control of machine tools. Numerical control, as this is called, has so radically changed the machining process that machine tool manufacturers are now producing numerically controlled machine tools almost exclusively; manufacturing companies have been forced to either reorganize manufacturing departments or else redefine departmental functions, and companies and jobs have been created involving skills that didn't exist fifteen years ago.

Just within the past two or three years, another computer-derived technology has emerged which has all the earmarks of an equally revolutionary impact. Identified in the trade journals as computer-aided design and manufacturing (CADM for short), it is just beginning to alter radically the processes of design and manufacturing in industries ranging from automobiles and aircraft to textiles and transistors.

Ten years ago I myself was deeply involved in the transition to numerical control; I saw that, quite apart from technical questions, it raised profound problems of administration and management. Now I see many of the same organizational problems reappearing in this shift to CADM—and I am concerned that few managers are prepared to deal with them. Numerical control produced upheaval in companies where it was introduced, and CADM could produce even more, especially if the mistakes we made then are repeated!

Computer-aided design and manufacture is based upon two essential elements. The first is the technical data base, by which I mean the files of technical design data which must be captured and then stored within the computer for subsequent processing. The second is the graphical input/output terminal. This device, which has been commercially available only since 1965, consists of a cathode ray tube where information from the data base is displayed either as text or as geometric shapes. The designer manipulates this information by means of a keyboard or a light pen, a light-sensitive device he applies directly to the display in order to modify it.

In order to make clear how such a system can be used, I will describe briefly, and with some oversimplification, how one computer manufacturer uses it in his own design work right now. His data base comprises files of information describing the standard circuits and components that are available to his design engineers, the subassemblies and assemblies of these which make up a specific computer product, all the interconnections, and, finally, the physical arrangement of the components on the circuit boards and racks. Not all the information needed is always available in the files, of course. At first, most of the files are empty, except possibly for the file of standard circuits and components. Each design engineer, using the graphic terminal, can interrogate this file to learn what components are available to him, and the characteristics of the circuits to which his design must interface. Using his light pen and keyboard, the designer can then assemble basic components into a complex circuit design that will be shown on the face of the tube. Next he can command the computer to perform various analyses of the probable performance of this tentative design and display the results. In this way he can change the design and re-analyze it until the circuit meets all the design criteria. Finally he can command the computer to store his design in the appropriate file for reference by designers of interfacing circuits.

At the appropriate time, the information in the files is processed to produce tapes which control the automatic wire-wrap or component insertion machines that actually manufacture the circuit boards. The file information is also processed into tapes to control the automatic inspection machines that test the final circuit boards. It is also possible to use the file information to produce schematic diagrams for the maintenance personnel who must diagnose and repair the installed computer hardware.

This example, which is probably the most advanced use of CADM to date, points up the broad applicability of this new technology to the computer and electronics industry in general. Indeed, IBM, Norden, and others have been using such systems for several years now.

But CADM is by no means limited to electronic circuits. In my industry, aircraft manufacturing, these techniques are being used to perform in hours a variety of chores that would previously have taken days. For example, at McDonnell-Douglas and Pratt and Whitney, fitting curves of observed data to empirical formulas is done by displaying the points to be fitted, along with the mathematical formula selected to go through the points. An engineer can immediately see how well the formula fits the points, and can change the formula again and again until he gets the best fit. At Boeing and at North American Rockwell, CADM systems are used for conditioning flight test data. This involves displaying plots of the data transmitted from satellites and missiles, and using the system to eliminate spurious data, or to fill in gaps caused by short transmission failures.

More sophisticated systems perform simulations. At Boeing, for instance, extensive use has been made of the computer in simulating aircraft landings on runways and carrier decks and displaying the result from the pilot's viewpoint as an aid to the designers in evaluating their designs. In logistics, the process of mounting and dismounting engines in an aircraft has been simulated and displayed in order to evaluate and anticipate field maintenance problems. Lockheed is using CADM for mechanical design and drafting—as is General Motors with its DAC-I system, which also assists the automobile styling effort.

Experimental system have also been developed to aid in the design of freeway interchanges, in the design and internal layout of buildings, and, more recently, in abstract design of pure forms and colors.

CADM is presently at about the same stage as numerical control was in 1957, just after the first production machine tools with tape control became available. For instance,

> In those days the electronic control systems were considered too expensive; at present graphic terminals are also expensive.

In those days the boss knew that numerical control was all right for milling machines but he wasn't sure what else it was good for; today CADM is obviously of value to graphical problems such as drafting and curve fitting, but in other areas it is less so.

In those days many people thought that numerical control simply involved making a tape and keeping everything else the same; today not even the experts are able to project just what the impact of CADM will be.

This uncertainty over the precise nature of the changes CADM will bring about is, of course, at the root of the difficulty facing any management that is contemplating getting into this new technology. Moreover, though there are strong parallels with the technical and management problems that occurred in introducing numerical control, in many cases the senior managements that struggled with those problems are no longer around. Today new managers face problems which, in many cases, are the same ones all over again.

As I look back at numerical control it seems clear that many of the difficulties came about because (1) it was a new technology, (2) it was expensive to implement, and (3) its capabilities were not well defined.

As a result it involved a very high degree of risk.

Then—as now—the crucial factor in reducing this risk was a high degree of top management commitment. By this I mean that the chief operating officers of the company cannot merely give their approval to entering a new area; they must become actively involved and, to some extent, technically knowledgeable as to just what it is they are getting involved with.

Too often a company will simply install some new equipment and tell a lower echelon manager that it's his job to get the equipment running. But from personal experience I know that frequently he simply cannot do the job by himself. The impact of a technology like numerical control is felt across many departmental lines, and a lone manager lacks the authority to make the necessary changes in the existing corporate structure. He may not command the necessary manpower or money.

Worse the middle manager may not be able to choose intelligently among the many possible methods of implementation—because management has not been able to define its own objectives and to transmit them, clearly stated, to people down the line.

As the relating of capabilities to corporate objectives is such a key point, let me get a bit more specific.

Suppose your major concern is to find cheaper ways of doing things. Certainly there are a few areas in which computer graphics are cheaper. On the other hand, a word of caution is advisable. Most of us who were involved in the early days of numerical control will tell you that it took three to five years of productive application before there was sufficient data to demonstrate cost effectiveness. There were always a few gee-whiz cases around to "prove" we were saving money, but there were also a few "Great Scott, what happened?" cases.

One of the difficulties is that companies do not always conduct their accounting in such a fashion that direct comparison is possible. I know of one company with a fairly well developed computer-aided drafting system. In one division, the system was demonstrated to be cheaper than manual methods. But the same system failed the cost-effectiveness test in another division, because that division accounted for its overhead costs differently.

This problem shows very plainly in numerical control manufacture. Here the preparation of tapes to control the tools replaces functions normally spread among manufacturing, planning, and tooling departments. In a company in which the individual parts are not end items, such as the parts of an aircraft, the accounting system rarely isolates the cost incurred by each of the departments involved relative to a particular part. To do so would be a difficult task, involving not only the direct costs of each department but also the pro-rated clerical, managerial, and overhead costs such as power, lights, and floor space. More important, management is likely to be concerned about the total costs of assemblies and end products rather than detailed costs of individual parts. It may turn out that the only way to isolate the cost of designing and manufacturing an individual part is to send it out to a sub-contractor.

In any event, CADM is at such an early stage of development that by any method of measurement, and in many specific applications, it is not yet cost-effective. If economy is your primary, short-range goal, it may well be that CADM is not for you—yet.

On the other hand, there are companies, notably in aerospace, that are primarily interested in acquiring greater capability—a better way of doing things. They may be interested in designing their products faster, or in producing better designs in the same amount of time. If a designer has a tool by which he can look at a thousand designs and find the best one, he can produce a better design than if he only examines a hundred in the same period of time. To take a specific example, an airplane designer able to use a CADM system that allows evaluating a thousand solutions instead of only a hundred should then be able to design an airplane that can fly farther, or faster, or higher, or be able to carry a bigger payload.

It is important to clearly distinguish the objective of greater capability from cost reduction. For instance, the airplane designer may well be spending more with CADM. In the past he probably had a fixed period of time to do his job—say four months. With a CADM system, he will produce a superior design, but he will still take his scheduled four months. Therefore the total cost of doing the job will be greater because of the cost of operating the CADM system.

So far, I have considered two possible objectives—cost effectiveness and greater capability. There are others, depending on the kind of company you have. It is up to management to decide what its objectives are and gauge them against the progress of a developing field. Numerical control developed first in aerospace, where the need was for greater capability. Then as the technology became cheaper and more powerful, the larger commercial establishments such as the automobile industry began moving in. And so the process went until the machines became so cheap that today even small machine shops are equipped with tape-controlled machine tools.

I expect CADM to follow roughly the same route. At present, much of the pioneering effort is being done in aerospace, with notable exceptions such as the developments by IBM, Control Data Corporation, and other computer manufacturers, and the DAC-I system developed at General Motors over the past several years and now being used for automobile design and styling. As the technology improves we can expect cheaper graphic terminals and computer software which, in turn, will further expand the use of CADM systems. Ultimately, with the continued lowering of costs and with further improvements in data communications, we may well find graphic terminals in thousands of small design and drafting shops across the country, all linked remotely to CADM computer centers.

Let me move on now by assuming that your management has spelled out its objectives and decided to move. Its next task is determining the organization required and defining people's responsibilities. In terms of avoiding disasters, this is where the payoff lies. I recall one company where management decided it was time to get into numerical control. Innocently assuming that their existing organization was up to the task, they ordered a complement of tape-controlled machine tools. Their industrial engineering people did their usual good job in arranging for the procurement, transportation, power, and installation. The production people budgeted carefully for operators. The machines were in the process of being installed before it dawned on anyone that no arrangements had been made for equipment to punch tapes or to duplicate tapes. Nor had any provision been made for people who knew the computer well enough to provide the necessary programs for that end of the system. Finally, the problem of people to handle the maintenance of the sophisticated electronic control systems had not been foreseen. The result—machine tools standing idle until the problems could be solved.

Another horrible lack of preplanning occurred in a company that had successfully installed numerical control equipment. By doing this, they put a tool in the hands of the manufacturing people so powerful that one day several hundred parts were manufactured and inspected before word came down that the design had been changed. Management had not realized that the pacing of design and manufacturing had previously been in balance and that numerical control had radically upset this balance.

The real problem, of course, in these two cases was that management had not recognized that numerical control was something really different, and as a result they did a bad job of organizing for the new technology. I can hardly over-emphasize the need for recognizing that CADM is not just an improved way of doing what your company does now.

The trap is an easy one to fall into, of course. Before numerical control, manufacturing companies were pretty strictly compartmentalized. Engineering's job was to produce a set of drawings adequately defining a product, and that was the end of their responsibility. Production people took the drawings and started to plan how they would manufacture the product. But with numerical control, the design of a part is affected by the fact that it is going to be machined on a numerically controlled tool. Therefore, the designer should have this information before he does his job—information which used to be generated by the production people after the fact.

Another problem with numerical control arose because programming of tapes for part production is necessarily accomplished with the aid of computers. Therefore, the production cycle is dependent on the operations of the computer center, where the part programming jobs must compete with other jobs from the engineering and business departments for time on the computer. Thus the production schedule can be drastically affected by computing priorities and downtime, factors over which production may have little or no control.

This type of problem—a blurring, if not elimination, of interdepartmental boundaries—will certainly be magnified with the installation of CADM systems. With engineering, tooling, and manufacturing all inputting to and retrieving from the same master files, virtually the entire organization will be tied together into one intimate process; everyone will become extremely dependent on everyone else, and functional responsibilities will be difficult to define and enforce.

How to cope with this new togetherness? A couple of approaches come to mind from the early days of numerical control. One, of course, was to set up a committee where the line managers from all the departments involved got together on a regular basis to discuss procedural problems. This approach turned out to be effective—indeed essential—in dealing with the routine of keeping communications flowing and getting each department to recognize the other's problems. Often, problems could be resolved on the spot.

But the committee approach can be very ineffective when action must be taken. As, for example, when two people of equal authority could not resolve their differences over whether the machine tools were to get their regular preventive maintenance during regular work shifts (the most convenient for the maintenance department) or on weekends (the best for the production department). As a result, most companies with a committee approach found they had to select a coordinator—an individual with the responsibility and authority to resolve such problems as they arose. Such an individual is, of course, hard to find. He has to have an awfully thick skin and a lot of knowledge relative to what everybody's capabilities and duties are, and how to relate these to the company's objectives.

Another approach is the job shop approach, where you essentially eliminate compartments and create a company within a company. I know of a couple of companies that dealt successfully with numerical control by creating a single department and putting all the necessary design, tooling, manufacturing, and electronic maintenance people in one room. By keeping the operation relatively small, communications were very tight, and no one had to walk for more than ten desks to talk with anyone else.

Of course, this approach was used mainly by companies that were relatively small in the first place; its applicability to a large company is questionable, although for CADM management should recognize that a single system need apply to only one specific product. Thus, the organization of the company into product lines, each equipped with its own individualized CADM system can certainly offer a practical approach in some cases.

Each company will have to base the approach it takes on the nature of its product line, its existing organization, and above all, the people involved.

It is clear that the very highest priority must be given to adequate preplanning by top management. For when it decides to install CADM, management is taking a grave risk—it could lose control over the entire manufacturing operation. To avoid that is the role of preplanning.

Management must do more than design an organizational structure where all the parts interlock properly. Quality control checks will have to be built in to assure both the timeliness and accuracy of the technical data base, and there must be feedback loops to remove errors from the data. Here's what I mean:

Presently many engineering department incorporate checking groups to assure the quality of drawings and design. Often, the checkers have the responsibility for making certain that drawings conform to specifications, are intelligible to manufacturing, and that the design is consistent; that is, parts designed to fit together will actually fit when made to the drawing specificaions. If the drawings fail these tests, the checkers take them back to the designer, thus closing a feedback loop.

Similarly, the machinist in a traditional machine shop was always providing feedback. In the process of machining a part he would set his machine's feed and speed rates to those determined by the planners. These rates would often turn out to be incorrect, but he could detect this—and correct for the errors—because he was constantly watching the color of the metal chips or gauging their temperature by spitting on them.

Numerical control eliminated the machinist's feedback loop. Here everything was planned and incorporated into the tape. This led to trouble, since the machine tool did exactly what it was programmed to do—right or wrong, the garbage in, garbage out principle. This problem was solved by developing electro-mechanical equipment that could sense the performance of the cutter during the machining operation and automatically adjust the feed and speed to an optimum condition, thus providing an automatic feedback loop.

A computer-aided design system with a centralized data base will produce a strong dependence on the stored data. The designers themselves may introduce errors into the data base, and these errors can propagate all the way along to the finished parts. It may even happen that the inspector, using data base information, inspects the same error into the part. It may only be when the parts don't mate with one another that the error will be uncovered.

The problem of error propagation and control becomes even more difficult in the case of a data base so large it must be decentralized into several smaller files. (The cost of processing a file increases geometrically with the size of the file.)

Consider, for example, the generation of an automated wire-list system. This is essentially a list of all the wires associated with an airplane or missile, together with information regarding the connectors on the ends of each wire. These lists are prepared by the electrical engineers who use them to trace signal paths while checking performance of the electrical system. But the information is useful to many other people also. Purchasing can use it to buy components; assembly people use it to build and install the wiring harnesses; and field maintenance can use the information to diagnose faults in equipment already in the field. In all, six to ten separate reports are produced from one set of data.

A problem arises because each person requires the information in a different form, and so the data must be transformed for each individual report. In a system such as Apollo/Saturn, which has some two billion wires, these transformations are so voluminous they cannot practically be performed each time there is a change in the basic data. The solution is to maintain a separate file for each type of user with essentially duplicated information in each file, but arranged specifically for that user. Though this means more efficient processing of the files, it also requires a fairly sophisticated system for eliminating errors that have propagated throughout the files. Really large data bases comprise a problem which we are really only beginning to look at, with no very good solution apparent.

Nevertheless solutions to technical problems such as these are primarily a matter of pouring in enough resources—money, manpower, and equipment —for a sufficient length of time. Far more troublesome, in my opinion, are the people problems, which become most acute in the development of any new and different system of doing things.

I recall a number of problems from the early days of numerical control that centered around the part programmer. He was a new kind of specialist— he had a skill that hadn't existed a few years earlier, and, like the computer programmer, he spent a lot of time just looking at the ceiling as he went about his business of solving logic problems. This proved very unsettling for the shop foremen in this companies where—to assure better coordination—part programmers were placed in the manufacturing area. Manufacturing managements are imbued with the philosophy that if you aren't moving, you aren't working—probably correct in the case of machine tool operators—and the idea of not being able to tell whether a guy was working or asleep was sufficiently alien to the foreman that it soon seemed wiser to isolate the part programmers.

Another problem arose when a great many companies went into numerical control and thereby produced a severe shortage of people skilled in the new technology of part programming. CADM managements may well run into the same problem, particularly with respect to draftsmen. Since a CADM system produces drawings automatically, there is no place for the draftsman, per se—for the man whose only function is to commit the design to paper. However, one company I know is already training its draftsmen to utilize a computer-aided system for some of the difficult or time-consuming parts of drafting. For example, the draftsman may have the problem of distributing thirteen rivet holes along some arbitrary curve. This is not a particularly easy task for him, but the computer can whip them out in a hurry. This company has developed a computer-aided system where the draftsmen themselves can use the computer and an automatic drafting machine to get these difficult jobs done.

Should you decide to embark on a training program, you will have to be prepared for the fact that everyone may not be trainable in the required skill. For example, there are people who have great difficulty in learning to use the light pen on a cathode ray tube. They have some sort of difficulty in hand-to-eye coordination that makes them inefficient in lining up the light pen with the object they want to point at.

There is also the problem of resolving stereo pictures. There are several techniques for viewing these picture pairs and seeing them as three-dimensional objects—for example, the old-fashioned stereopticon our grandmothers used, or the little View Master sets we give our children. But some people apparently cannot resolve these pictures into three-dimensional objects. Such people are obviously unsuited for work on stereoptical systems such as those used for aerial mapping. This kind of problem is one which must be anticipated by management, since it can so drastically affect the staffing of a new system.

You must also anticipate that management may not have total control over its personnel situation. When numerical control came in it became evident that the standard electrical maintenance people lacked the basic knowledge and skill in electronics to do a proper job of maintenance and repair. We needed a person whose skills lay somewhere between those of an electrical maintenance man and those of an electronics engineer. As a result, several companies set about providing job classifications and pay scales that would provide an incentive for electrical technicians to learn to become electronics technicians. Unfortunately, some companies got hung up in negotiations with the unions. The unions seemed to feel that all the electrical technicians should simply be trained up and given an across the board raise. The various companies solved the problem in various ways, eventually, but while the lengthy negotiations went on the maintenance of the machine tools became a critical problem.

A question that was always of great concern to us in numerical control was the emotional reaction of people to change. After all, the word sabotage comes from the early days of the industrial revolution when the Dutch workers dropped their wooden shoes into the machines to break them down. In the paper processing industry, vigorous and sometimes physical battles have taken place over the introduction of machines which directly result in people being laid off.

In numerical control, somewhat to everyone's surprise, most companies found the machine tool operators eager to get involved. The machines were new and oftentimes the man who had been running some old clunker for twenty years welcomed the opportunity to get a new machine and do something different. In many shops, there got to be some competition as to who would get the choice spots, namely running the numerically controlled machines.

I believe this enthusiasm arose because numerical control never really eliminated people, but in fact created jobs. Since the machines were better, more things were done. Since the machines were somewhat more expensive, management felt more pressure to keep them busy with work. This often meant going from a one-shift operation to two or three shifts, thus creating more jobs for the operators.

I expect the same thing to happen in computer-aided design. The company that successfully develops a better, faster, or cheaper design system will naturally capture more design jobs. It will be up to management to convince themselves of this, and transmit this conviction down the line.

In any case, it is important that management know its people and anticipate their reaction to this new technology. This is most important at the beginning, when serious consideration should be given to staffing the operation with the people who are the most capable and the most amenable to change. When we started numerical control, we anticipated that eventually the operators would be in relatively low job classifications, since they wouldn't really have much to do other than handle the parts and blanks, mount tapes, and push the buttons. We did not feel that we needed a master machinist, because the whole operation would be preplanned, and programmed into tape. But we soon discovered that because the whole technology was new, and a high proportion of errors likely to occur, we were better off initially to put our very best people at every station. The master machinist operating the machine tool could quickly spot badly programmed feeds and speeds from the fact that the chips were too blue or the smoke too heavy. He could sense the kind of error that might result in the machine damaging itself and shut it down before this happened. Until the whole operation smoothed out, we had to utilize people capable and flexible enough to not only detect things going wrong, but also offer constructive ideas as to how to eliminate the problems. Afterwards we moved these people onto other developmental programs—which was no problem for our company but could be one for a small firm.

The same thing will probably be true for computer-aided design. Consider for example, the case of drafting on a cathode ray tube. You should have one of your very best designers involved, so that he can tell you when things are going wrong, help isolate the specific problem, and constructively work with the computer programmers in revising the system.

Further, the image principle applies to the first-line management. For example, it is not an easy job to manage programmers, who in many ways resemble research people in that they are often highly intelligent and creative, and have little patience with bureaucracy. Competent programmers are in short supply, they are extremely mobile, and turnover is high. Hence your CADM managers must also be first-rate.

In considering CADM, management must also pay close attention to the economics. Some time ago, I developed a set of economic factors—using part programming as a case study—that is still pertinent. The study was intended to determine the cost of operating an already developed computer-aided system for producing tapes for numerically controlled machine tools.

A range of hardware was considered, from a minimal system with a single graphic terminal sharing a computer with other systems on a single-shift basis, to a monster system with twenty-four graphic terminals on a single, dedicated computer, running around the clock. The cost of operating the minimal system, including the terminal hardware, computer time, and one system programmer was $7,250 per month. The monthly cost of the large system came to nearly $150,000. Unit costs, represented by the cost per terminal hour, amounted to $100 for the minimal system, and about $25 on the other.

By comparing these costs with the present cost of part programming, I determined that productivity would have to be improved by a factor of from 4 to 6 if the CADM system was to prove cost effective. Development of the system was estimated at $250,000.

Though these figures should not be applied to situations other than the rigid environment of the case study, they do illustrate the level of costs that could be involved in a high-capability CADM system. It can certainly be anticipated that systems for other purposes and not involving the currently expensive graphic equipment could easily cost much less. On the other hand, we must realize that the things we're doing today are going to look small ten years from now. I can remember when it was a traumatic experience to decide whether to get the $40-per-month key punch and save money, or get the $50-per-month key punch with its greater capability. Today it's difficult to keep track of how many key punches we have at twice the price. I also remember talking to a university administrator who remarked wistfully that today's bill for cards and computer paper is larger than his entire R&D budget twenty years ago.

Nevertheless, it is clear that development costs can be high. Further, at the present time very few systems are available off-the-shelf, so that nearly everyone getting into CADM now will have to face a development phase. In general, the make-or-buy decision is easy since there isn't much available to buy.

Another factor leading to high development costs is the speed at which the technology is advancing, leading to rapid obsolescence. The system you put together today is very likely to be out of date in one or two years. This is not entirely bad, since the first primitive system may allow you to better define system objectives and give your organization time to adjust.

There were some companies in the early days of numerical control who approached the problem exactly this way. They were confident that the APT system, pioneered at MIT and being developed at that time under the Aerospace Industry Association's sponsorshop, would provide tremendous capability for computer-aided part programming. Since APT appeared to be one or two years away at the time, some of these companies elected to develop interim systems. These were designed fairly cheaply to bridge the gap until APT would be ready for production use.

These companies found it most effective in the long run to develop systems with the intent of throwing them away at a later time with no regrets. Further, the experience gained with the interim systems had a profound influence on the development of APT, helping to make it the efficient tool it is today.

In the same way, we can expect CADM technology to change rapidly over the next few years. Graphic consoles will become cheaper and more versatile than they are now. Equipment will be designed for specific tasks as various systems come into more widespread use. Software houses will become involved in developing systems for sale or lease as off-the-shelf products.

But management has to decide whether it should break new ground or not. It may be that, like the interim part programming systems, the pioneering system will provide a better way of doing the job. This in itself gives sufficient justification for being a pioneer. In other cases, it may be wiser not to take the hard knocks involved and simply wait until the technology has matured to the point where you can buy the system needed to do the job.

In making your decision, there are two dangers to be avoided. One is in being too ambitious. I feel that some companies today are spending too much money which they may never get back. They are not edging their way in so that they can reverse field and take off in a new direction easily. The opposite problem is that of depending too much on the cost-effectiveness criterion. The company with this approach may not be just stifling its own progress—it may never even get into this new technology. The concrete evidence it keeps asking for may never exist.

Regardless of the direction you take, you must have knowledgeable people in the line between management at the top and the operators at the bottom, and you have to trust them a lot. Then, as long as you have good communication from the top as to what the company is trying to do, and what technology is needed to do the job, you should be in good shape.

Until the next revolutionizing technology anyway.

January 1970

11

What Happens When You Invade a High-Technology Market?

You don't have to invent a technology to bite off a piece of the action. But you do run into a different set of problems than when invading a less esoteric market. IBM was already one of the leaders when Honeywell decided to go into the computer business.

Evan Herbert tells how the newcomer went about organizing a successful market strategy and how these efforts forced changes in the company's approach to technology itself.

Mr. Herbert is a Senior Editor of *Innovation* Magazine.

The amiable animals nibbling at the IBM computer in Honeywell's advertisements are part of corporate strategy. They are trying to bite off a profitable piece of the approximately $4 billion computer and information processing services market.

Practically every company that tries to capitalize on a brand-new and complicated technology faces a characteristic set of problems. It has to get at a marketplace that hardly understands the technology being offered—and this marketing requirement forces changes inside the company itself, changes that affect the way it designs its products and provides for their technical support.

When a radical new technology first develops, about the only people who really understand it are technical persons. Yet these often are the people least able to communicate and interpret what the technology offers, either within their own companies or to the marketplace. High technology is so expensive that decisions to buy it necessarily rest on the shoulders of senior officers least able to understand what a technical man tells them.

The communications chasm between technology developer and ultimate customer yawns wide and deep. Building bridges across it is neither simple nor inexpensive, even when you're an acknowledged leader. It becomes even more difficult when you come a little late to a field and have to bite off a piece of what the leader already has.

For the purposes of this essay, the technology is computer data processing and the company is Honeywell, Inc. Honeywell decided to enter a business in which IBM was a front-runner already far ahead of the pack. The spectrum of techniques Honeywell used to organize itself to market a technology it had never been in has now placed it second in cumulative value of computers installed for this market.

In advertisements, their animals don't look so voracious. They're "nice" animals because company policy decrees it. But they do mean business, and so does Honeywell.

For Honeywell is the only one of the eight substantial computer companies that has chosen to meet IBM head on. IBM has 70% of the business, and the other companies, somewhat bloodied by battle with the giant, now seem content to concentrate on specific areas of application where they can compete.

Competing, in the marketing of high technology, is much more than fielding more salesmen, more than mounting stronger advertising campaigns. In the case of computers, marketing activities cannot simply begin after a new product has been created; ideally they shape the product by being attuned to the environment it eventually will enter. Often the influence of a strong marketing organization effectively reaches all the way into the setting of directions for basic research. Unlike many consumer products, where money is made as soon as goods go out the factory door, computers are intended to have a long, useful life.

Few computer systems are bought outright by the ultimate users. Computers are more often outgrown than obsolescent, so computer systems usually are leased, either from the manufacturer or from a growing array of leasing companies acting as middlemen.

Whatever way a system is acquired by a user, somehow he must end up not only with complex machinery, offhandedly referred to as hardware by suave-looking computer salesmen in Brooks Brothers suits, but also with a spectrum of programs called software. Add service and maintenance, plus a skilled staff educated in the peculiarities of the hardware-software combination, and a computer system hardly comes cheap, nor is the investment in it short-term.

Indeed, the acquisition or upgrading of a computer system these days is approached by most prospective customers as cautiously as marriage to a demanding woman with expensive tastes and a ready-made family. Computer marketing people privately concede that this analogy may be valid, but one of their jobs is to help a customer find true happiness with a computer. They say it's better to think of acquiring a computer system as like getting a puppy—there is a lot of growing up together. Besides, the customer can always swap his smart miniature poodle for an even harder-working St. Bernard.

No matter which analogy is correct, the trouble is that "the customer" is not just a single describable person. When it comes to getting a computer, a lot of people get into the act, even though only one of them signs the contract. The "customers" in any organization range from technically-oriented management through corporate financial watchdogs and on up to the chief executive officer and even the chairman of the board.

All have to be sold, in varying degrees, on the merits and utility of a technological product hardly as simple as a puppy—and possibly as complicated as a woman. Often it's a matter of educating each of the different key people about the technology so they acquire a common basis for understanding how the product may solve their problems. In the process, they often sell one another.

In bringing their high-technology products to the marketplace, computer companies are discovering that they are tied-in to two *kinds* of decisions made long before any contracts get signed.

First, they must set up the system architecture—the functional design that precedes hardware. This means anticipating customer needs three to four years into the future, even though few customers can predict or describe them.

Yet the design and engineering of systems to meet these needs take up to three years before the hardware *and* the software ever get delivered to the first customers on the waiting list. Meanwhile, a whole stream of peripheral products—printers, graphic terminals, tape-readers, disk memories, etc.— may enter the marketplace from many small companies. The lead time on these is often much shorter than on complete computer systems. Yet such equipment may become meaningful to the system hardware and software designs already in progress, for either the manufacturer or an ultimate user may wish to incorporate them in a system.

The other kind of decision involves a more subtle problem than advanced design: how to help customers themselves to grow along with the development of a technology so radical it might well cause them to scrap their investment. While the explosive growth of the computer field is easily attributable to technological innovation, the sudden wholesale introduction of drastically different technology—as in the shifts from vacuum tubes to transistors to integrated circuitry—creates generation gaps that most customers are ill-prepared to bridge.

Customers don't necessarily want to understand the newer technologies that create these generation gaps, but they do care about the disruptions created by technological advances when the acquisition of better equipment also means expensive reprogramming and retraining.

For the manufacturer, the real money to be made comes from getting his customers to upgrade smoothly to larger and larger systems. Naturally one's own customers are the best prospects for upgrading; they are already familiar with the equipment line and have a substantial investment in programs and personnel.

There are also the competition's customers to be lured, and the key to attracting them is compatibility with programs they are already using, even though they may have been developed by the competition.

In the beginning though, designers knew more about technologies than about customers' problems. The early key men in the computer field—only about 19 years old right now—were essentially technical people, too. Except when they were talking to other technical people from the very same field, there was a mystique about what they did and what their products could do for a customer.

From the customer's side, even though his technical men sometimes participated in the early decisions to acquire a computer, most computers were chosen more on faith than on understanding. And when in doubt, there was always the safe out of going to the biggest manufacturer in the computer business. No buyer would need to feel guilty if it turned out that not even IBM could quite make things work.

Even today, there is an enormous gap between what a customer's technical man thinks he wants in a computer system and what the top management man thinks he is going to get. Yet both expectations count—which is one reason why a marketing organization for high technology gets as involved in the directions of basic research and in product planning as it does in actual sales.

Along about 1955, Honeywell reasoned that it could significantly strengthen its position in the automatic controls business if it had computers to offer, though few people in that company knew much about the computer business.

Some 224 large-scale computers had already been announced by IBM, RCA, and Remington Rand-Univac. Figuring that the market would be perhaps 300 such machines and feeling certain that they might sell even a dozen machines costing over $1 million a piece, Honeywell formed a joint venture with Raytheon, which already had experience in the development of Whirlwind and Raydac. Honeywell turned out to be as wrong as everyone else, conservative or bold. Now, just fourteen years later there are 1,300 large-scale machines in the world each worth over $1 million, and their capacity makes the early big ones look puny by comparison.

As Honeywell-Raytheon started out with a company called Datamatic, the men who best understood what a computer could do were recruited among Honeywell's industrial and electrical engineers. These engineers assumed that the potential customer was as technically sophisticated as they were. It was a costly assumption. In retrospect, it seems possible that twice as many computers might have been sold then if that assumption had not been made. Then as now, the key computer buyers are relatively unsophisticated, technically. Buying decisions are made at the level of president, vice president, controller—even though today these officers often have sophisticated technical staff and consultants. Any given computer offers a spectrum of capability which has to be matched up with the entire spectrum of a company's potential application before reasonable justification to buy can be achieved, and that's not a job for specialists.

As it entered the 1960s, Honeywell marketers reasoned that IBM had already convinced a lot of buyers that computers indeed were useful. Thousands of 1401 computers were already in use and producing good revenue for IBM, which made it seem likely that IBM wouldn't want to replace them too soon. If Honeywell could equip a replacement machine specifically designed to deliver better performance and price, those 1401's would be sitting ducks.

The design would have to make it simple for the customer to convert those systems without a lot of supporting services from Honeywell or a lot of trouble for the customer. In short, the customer, already sold on computers, had to be convinced that he could buy even more performance for less, and he had to believe that he was not wedded forever to the equipment and programs of one supplier.

Perhaps the key factor in the selling of the Honeywell 200 as a replacement for the IBM 1401 was a concept called The Liberator. In essence, this was a software technique by which programs could be translated by the new equipment into a form that would play correctly on it. At that time, IBM had an emulation technique; it made a particular piece of new equipment seem internally like the old one for which a program had originally been written. Eventually, to make the program most effective on the new machine, it would have to be rewritten. Translation has a lot more appeal on the market because programming skills are always in short supply.

There was probably another, more subtle factor at work behind the concept of The Liberator—the basic need for a fresh, strong source in the marketplace. While IBM had enjoyed its phenomenal success, a lot of other companies, such as Bendix, North American, Hughes, and Philco, were falling out of the computer business after agonizing and costly experiences.

Honeywell experienced its own share of technical troubles, but by 1960, customers had become experienced enough to understand what it took to build a complex product like a computer. Honeywell banked on the new sophistication of the signed-up customers by taking them into its confidence whenever technical problems threatened to cause a slippage in delivery. Honeywell figured that once its customers understood that slippages were caused by attempts to do what had never been done before—faster-than-ever memories, new concepts in compilers—most of the would be willing to go along with the delays. They played it straight.

However, the explanation of the slippages was not simply left up to the sales and engineering departments. Honeywell's president went out to visit most of the early customers who might be hurt by delivery delays, and he made it clear that the company had no intentions of staying long in the pattern of late deliveries.

Many of the customers approached this way were even willing to sit down in the factory to listen to the problems and decide whether they seemed susceptible to solution in the then state of the art. A lot of customers hired outside consultants, and Honeywell brought them into the factory, too.

The subsequent record of shipments became a strong selling point. While rival salesmen were still talking about forthcoming machines, Honeywell began to ship pretty close to schedule. For example, the production shipment cycle of its Model 200 computer went from about 2 machines per month to 60 machines per month in just nine months.

The Model 200 involved another major strategy decision, for it was to become the first of a *family* of machines. In 1962, there were no families of machines compatible with each other in the sense that to enlarge one's system one simply could add larger central processors or more or faster peripheral gear.

At that time, the computer business was in such a state of great technological flux that compatibility was difficult to engineer.

Up to about 1958, computers had been based on vacuum tube technology, and though numbers of machines were often produced bearing the same model number, they weren't particularly standard, as one could discover by comparing machines of early and late serial numbers.

Then the solid state era began, founded on transistor circuitry. But even this wholesale change in technology did not immediately breed families within the new generation of computers. At best, there would be several models of the same *size* machine and these might be compatible.

With the advent of hybrid circuits and integrated circuitry in 1963, the computer field launched its third generation of computers. IBM announced its 360 system family all at once, though on a progressive delivery schedule, and Honeywell made it clear that the 200 was merely the first of a family. With that technological commitment, Honeywell had also committed itself to compete in all segments of the market, ranging from relatively simple applications of small machines up through complex problems or massive information processing tasks on very large machines.

This entry into a broader market was intentional. Looking back on it now from a perspective of nine more machines added to the family since the 200 was first introduced, it is possible to see how the strategy worked and the kinds of supporting organizations required to pull it off.

Companies just starting to use a computer tend to start modestly. Even if they have had punch-card equipment, the step up to a real computer is still awesomely high. The baby machine of the computer family may cost $2,200 per month, and it ought to be kept productive.

To help the customer with his first computer, Honeywell deliberately puts its newer salesmen on such an account. They're experienced enough to be of great service, but still fresh enough to computers to have rapport with a customer somewhat uncomfortable in the presence of new technology. Their goal is to make the customer so comfortable that he is anxious to try new things on his computer. The more he becomes imbeded in new applications, or in more sophisticated ways of doing things, the more likely he is to upgrade his equipment.

At first, salesmen were bird-dogs, backed up by engineers. In the early 1960s, the marketing strategy was to sell hardware performance and cost. There was a lot of razzle-dazzle about the number of bits processed, or the number of computations per microsecond, or how much could be held in computer memory. To most people, the numbers were impressive but hardly meaningful.

But once a computer user had had a significant taste of basic applications, he became wise enough to look at systems technology rather than at hardware technology. So the best strategy in the long run was to sell systems capability, and that meant the salesman had to know the problems of an industry well enough to deal with the sophisticated requirements of a manufacturer or an educator or a hospital administrator.

Honeywell, like its rivals, began to establish marketing operations which specialized in particular types of customers. In Chicago, for example, a whole branch office was staffed just for manufacturing operations. Not only did the salesmen there understand manufacturing because they had been in some phase of it, but they had the technical support of systems specialists. These were backed up with hardware-software technological specialists, people who knew computer operating systems well or who knew communications systems.

Above all, it is the systems specialist who has to provide continuity with the customer, because the best of salesmen are a fleeting sort. If they are any good at all, they are as ambitious as hell, and despite sales incentives, they won't stay with a customer long enough to thoroughly lay the groundwork for upgrading. The capture and continued development of the customer becomes an educational process.

The trick is to capture the potential customer's attention long enough to give him full measure of all the information and subtle assurances that go into his decisions to buy, replace, or upgrade. Computers don't fit easily into briefcases, so the interested customer has to make a pilgrimage to some center where a working model sits enthroned. Few buyers can afford the time to visit every computer company, so there is a natural tendency to gravitate toward the offerings of the six or seven biggest companies in the field.

IBM has so dominated the computer scene for years that "IBM machine" is often used as a generic term for computer. Unfortunately for Honeywell, its name was practically a household word, too—for thermostats. But being recognizable as an institution associated with regulating furnaces hardly gained it acceptance as a purveyor of the newest of high technology.

Honeywell tried another tack, suggesting in advertisements that "There's a little bit of chicken in all of us." Aiming at a computer buyer's tendency to play it safe by sticking with IBM, the copy tried to give him the gumption to look around. But it became clear that this strategy merely lumped Honeywell with all the other computer companies that might offer some alternative to what IBM was selling.

What was really needed was a way for the company to become *the* automatic second consideration whenever someone shopped for a computer. Years ago Chrysler Corporation had pulled off this strategy when it was trying to get its Plymouth into the same league as Chevrolet and Ford, urging everyone to "Look at all three."

Honeywell wanted to be thought of as the *other* computer company, and it set out to call itself just that with a heavy advertising and promotion campaign. In a left-handed way, Honeywell was excluding all other competition but IBM from consideration, either by the customer or by its own sales force.

Such a tough-minded approach implies a hard sell which most people resist. Yet in this case the goal was not merely to invite comparison, but to be friendly enough about it to bring the customer in for a visit. Hence the decree that the animals constructed from computer parts and used with phenomenal success to attract attention to advertisements for the other computer company, have to be nice. In fact, everything about the marketing effort had to be nice, nice enough to lure a prospective customer's ultimate signer of major contracts, the chairman of the board or president or vice president or controller, in to Honeywell's home base for several days' visit.

Several days? It does take at least that long to teach enough computer technology in general to let such people appreciate Honeywell in particular. But, even so, several days of a chief executive's time?

Few top executives have been exposed to computers as they come up the ranks, and Honeywell guessed rightly that many would welcome a management overview. They could get to learn enough fundamentals about programming, system design, or communications not to be buffaloed by their own staffs.

In the two and a half years that Honeywell has run executive-education seminars, more than 2,000 company officers have traveled to them at their own expense and the program is usually booked four months in advance. Most of the attendees are business school graduates and not technically oriented.

Still, the ratio of new prospects to actual Honeywell customers in each class is kept to about 3 in 25. The customers may be ripe for upgrading. All soon discover from each other that having problems is simply part of having computer technology, no matter who the supplier is. Yet they find considerable assurance of Honeywell's ability to solve problems as they work side by side with the instructors. Lectures and demonstrations are only one part of a daily agenda that begins at 8:15 A.M. and continues into mid-evening. The attendees get *involved* by actually writing simple programs and running them on a computer and by participating in computer-centered management simulation games.

Quite deliberately, the seminar is held in the warm, noncommercial environs of a rambling village inn, circa 1716, in historic Concord, Mass. Even the computer terminals don't seem intrusive. The bar, crammed with memorabilia, serves a "Flintlock," and one suspects that this may be a weapon aimed a bit southwest toward a rival's seminar facilities where no alcohol is permitted on company property.

Long before the visitor arrives, a full-time guest relations staff has spent time with the salesman on the account to find out exactly why the visitor is coming. Perhaps he is anxious to talk about a specific application. Then technical people with specific talents have to be made available to him while he is there. Or perhaps he is about to upgrade to a larger system and wants to talk to the array of people who will help him to make the transition.

When the seminar attendees reach the demonstration environment at Wellesley Hills, they encounter showmanship with computers in an ultra-modern setting. It is reminiscent of the theater in a pavilion of a world's fair. The lighting changes dramatically, the demonstrations are put on by polished professionals, information density is high but free of jargon, and when the lights finally come up at the end, the audience is invited to join the cast on-stage to mingle among the computers.

The seminar program is budgeted at about $200,000 per year. Three full-time educators arrange and operate the seminars, drawing upon other marketing and technical capabilities for support. In the first year of seminars, drawing upon other marketing and technical capabilities for support. In the first year of seminars at Honeywell, 13 large-scale system orders were attributable to their influence.

It remains a closely guarded secret just how much money for this sort of backup to sales goes into the overall marketing of computer technology. Rival companies reveal some numbers but never all of them, which may indicate just how important such support really is. Indeed, a current damage suit against IBM by Data Processing Financial & General Corp., which leases computers, complains that IBM lumps all services under a single price umbrella, making it difficult to compete. One part of the complaint alleges intimidation of users who might want to acquire competitive peripheral equipment by the threat that they'll lose the technical support given on the computer system as a whole. IBM's counter argument is that it does make available, without charge to equipment lessors, such support services as classroom instruction, assistance in the planning of installations, and maintenance of programming systems.

Clearly, few small companies in the computer field can expect to compete effectively against the resources of the large ones. Simply multiply Honeywell's 9 to 1 support ratio by the approximately 500 salesmen it has, and the payroll for marketing and its technical backup seems staggering. At Wellesley Hills headquarters alone, Honeywell has 175 educators whose task is teaching others to teach. There are another 130 educators out in the field who are taught at Wellesley how to teach, and *what* to teach to branch personnel, customers, and prospects. What to teach keeps changing, of course, as the technology changes and as the hardware and software for sale change.

Out of Honeywell's marketing education office comes a steady stream of packaged courses—programmed texts, videotapes, films, instructor's guides, and student's manuals. They are used in various combinations according to the nature of the class being taught. For example, a technical education group of about twenty within the marketing education office first checks out new software developed by the programming systems division of Honeywell. In a way, they support the systems analysts, the programmers, and the technical men by becoming surrogate for the customer. By seeing how these sophisticated programming packages play on the systems, they determine what problems various levels of customers may have in using them. At the same time, the technical educators must translate aspects of the packages into lay terms so they can train the teachers who will teach the customers.

In every case, there is a long chain of interpretive communication linking new technology, understood mostly by its developers, to ultimate users. This is particularly apparent in the activities of the information technology department. It receives an industry-oriented programming package—say for an application in drug wholesaling or one in automotive manufacturing—from an industry marketing group sensitive to the needs of the industrial area Honeywell is anxious to penetrate. The information technology department structures around each package a course which shows the customer how to run it *productively* on a computer system. This underscores the present marketing strategy—to sell the benefits of the system rather than the performance of the hardware.

Perhaps it may become easier to sell computer systems to the next generation of managers. Honeywell aims to educate potential managers now by offering three-month intensive postgraduate courses. The end products will be junior business executives, not technical managers alone, with strong backgrounds in data processing. Almost everything the budding executive then sees, as he rises through the ranks of management, he sees in the light of his training in data processing. Naturally, he sees many things in the light of the Honeywell systems on which he was trained. To some extent, this strategy may counter IBM's long practice of getting its systems onto university campuses, and even into high schools.

Perhaps the marketing group with the widest interfaces in Honeywell is product marketing. They can't follow the practice in consumer goods like soap, where there are product managers for every line complete with profit responsibility for a particular item. Product marketing of high technology, at least at Honeywell, reflects the requirements of the field into technical development activities.

First this group—a mixture of technical people with hardware or software experience, and ex-field salesmen who may be more business-trained—spends a lot of time in the marketplace trying to identify needs and interpret them to the technical people in engineering development.

For example, better forms of computer input-output are in demand and the product marketers may spend a lot of time with the technical people of companies that build optical character-recognition devices or microfilm equipment. They try to determine how these devices are being used now and how future computer systems would have to be designed to either offer such capabilities or tie in with existing capabilities.

They also try to determine ways to assure long life in the marketplace despite the rapid changes in technology.

As the product development progresses, the marketers act as an interface between the manufacturing group, the software specialists in the programming systems division, and the field locations. They are active in developing tests for the new product and coordinating such tests at selected field locations.

When the product begins to take on final form, product marketing assumes the task of communicating with two distinct audiences.

Naturally there are the potential customers, whom they now try to identify even more specifically. For them they must create an image of the product which will have an impact on such a market.

There is also an internal audience at Honeywell—the sales force. Clearly they need straightforward technical comparisons of the new product with any competing products. But the sales force also needs enthusiasm for the product as well as help in knowing what to look for when identifying someone who would need that product. So product marketing develops internal promotional material to help launch the product inside the company, and a follow-on campaign for both internal and external audiences to keep the product moving.

One of the by-products of being attuned to the external marketplace and internal technical capabilities sometimes turns out to be the repackaging of several existing products into a whole new combination which can be even more attractive in the marketplace than the discrete products. Technical people concede that this is all part of the maturing of high technology. In Honeywell's experience at becoming the *other* computer company, marketing people seem to make the technology grow up faster.

June 1970

12

The Power of Uncertainty in Organizations

In this chapter we discuss the relevance of information theory to organizations—and ways of handling uncertainties.

Arthur Myron Tribus is a top technical manager at Xerox Corporation and a former Assistant Secretary of Commerce for Science and Technology. He speculates that a little entropy may be good for you—it may actually prevent your organization from dying.

Every administrator faces some basic problems when, in order to get something done, he must move information through several echelons of an organization: information almost never flows as fast or as accurately as he desires, or brings the precise results he has in mind. All along the route there are surprises about the ways people get the message, absorb it, and either ignore it or act on it.

In my old area of the Department of Commerce (Science and Technology) where I served until the last days of 1970, there are about 15,000 people in bureaus and offices across the country. I had to communicate with them on a variety of policies, ideas, and facts. I also had to conduct my communications from an uncomfortable position—from the top down. A message from my office set ripples in motion, something like dropping pebbles into a large pond. It was very difficult for me to know ahead of time where the obstacles were that might break the concentric pattern. Also, it was difficult to know exactly where a signal might go. For example, I might run into someone in the hall who said, "I read that speech you made last week." Only then did I realize that somebody distributed my speech throughout the organization.

This condition is not peculiar to government. I've shared the same experience and feelings with other administrators when I worked in industry and education.

Over the years I've developed increased understanding of this condition, even if not full satisfaction. Partly my understanding is the result of experience; after 25 years as a research worker, professor, director of research, consultant in industry, member of boards of directors, college dean, and a government official, I have some expectation of how things operate. But in larger measure, it's due to my training as mathematician and engineer. Surprising as it may seem, I've learned a lot about organizations (and patience) from my work in science and technology.

I believe there are mathematical/theoretical reasons why organizational communications get gummed up or blocked, why feedback is poor or nil, and why on occasion everything in the system seems to go to pot. Some of these reasons derive from the same logical foundations as the basic laws of thermodynamics which for me are rooted in the work of Claude Shannon in information theory. The *concept of entropy* helps explain what goes haywire in the communications channels of an organization, especially a large one like mine was the Department of Commerce.

This is going to be a speculative excursion on the role that entropy plays in organizations, especially creative ones. It is possible to develop this subject in a mathematical style, complete with complex formulae. But in this chapter, I wish to concentrate on organizations, not on equations.

Let's begin with an experience shared by millions of people every day. To get to work—my very real, nontheoretical work responsibility—I have to drive from home to office. And back again at the end of each day.

As a commuting driver, I'm pretty lucky. My daily route is attractive. It starts in a relaxing stretch of countryside, runs south along the river, and then loops into town. Pleasant as the route is, however, I have the usual driver's complaint: Why can't all those other drivers behind all those other wheels conduct themselves the way I do—clearly, consistently, safely? To me, that means giving easily understood signals, not moving a vehicle erratically all over the road, and following the established rules. At least that's the way I think I drive.

How does driver behavior figure into the problem of moving information through an organization?

Well, think for a moment about a long line of cars jammed up at an intersection. All drivers are looking at the traffic signal, waiting for the light to change from red to green. What happens when the light finally changes?

The green signal should mean "go" for the first driver in line. For everyone else, it turns out to be another kind of signal. The green light is only a signal to pay attention to yet another signal: the behavior of the driver directly in front.

The cars do not all move at once. Each driver must make a set of decisions before starting. Has the car ahead started moving? If so, how much space should I leave between us in case he stops suddenly? What are those pedestrians going to do; don't those boobs see the light?

Each driver looks for signals other than colored lights to decide when it's appropriate to move ahead. So instead of 15 or 20 cars getting through an intersection before the light changes back to red, only six or eight make it.

In contrast, when an Army sergeant yells out, "Forward march," all the troops start to move at the same time. The signal is the same for all the men; no individual decisions needed.

Unfortunately for the nerves of administrators, the traffic situation is more familiar in most organizations. The sergeant-as-signal works only in the military. In most organizations messages don't often reach everyone at the same time. And even when they do, people don't always respond to the message; often they respond to the behavior they see around them as others receive the same message. The result, a lot less gets done than we think could be done by a "perfect" system.

A little consideration of this point leads to the conclusion that there's more to communication than putting a signal into a channel. The channel into which we put our information contains many nodes at which decisions are made. Therefore, if we want to understand the sources of delay, we must think of the people along the channel who make decisions based on our signals.

I'm not talking about those things we call "communications problems," where messages get garbled or don't reach the right place. The signals I refer to may be perceived very clearly. But even so, there is often a problem. The person who receives the signal often cannot act upon it, cannot make decisions, because he's also receiving a lot of other signals from other sources. The signals may be contradictory; information may be ambiguous or incomplete.

Soldiers in a formation don't have to make decisions. Each marching order has only one possible response. The soldiers don't have any alternatives. But in more complex, technological organizations—especially where routine things are already put on a computer—everyone who is there is supposed to make decisions.

So I want to look at organizations as decision networks; I want to examine decision-making problems.

Think of some everyday phrases you use in talking about the problems connected with getting information and ideas spread around your organization;

"There seems to be noise in the system—I can't get my message through."

"I wonder if we're operating on all channels."

"The situation is getting thick; nothing seems to move in this system."

We're used to this kind of language. It has specific meanings to us. Yet this vocabulary, which we freely apply to organizations, comes from other fields. In many cases, such phrases come as analogies from mathematical concepts in the information theory of Claude Shannon.

Specifically, Shannon's use of the word entropy is very significant in my speculation about what happens to messages in organizations.

Claude Shannon is a remarkable thinker, whose famous work was originally published in 1949. His theory was based on accumulated experience with mainly electrical systems, but communications theory has strongly affected many fields including psychology, medicine, economics, and branches of engineering. I recently reviewed a textbook of psychology based squarely on Shannon's work as the fundamental starting point.

Shannon was concerned with the communication-carrying capacity of electrical wires or other communications channels such as those used for radio. Given a channel of certain physical properties, how much information could you send?

He published his results in a paper called "A Mathematical Theory of Communication." (When the paper was republished a year later, in 1950, it appeared under the title, "The Mathematical Theory of Communication." Nothing else was changed, only the first word of the title.)

Shannon developed a new function that can be used to measure what is transmitted in communication. Von Neumann pointed out to him that this function was already used in thermodynamics: entropy. So Shannon called his new function entropy, too. He meant it as a synonym for the word uncertainty.

He took the very original approach of looking at the essence of the problem subjectively, not objectively. He looked at it from the perspective of the receiver's state of knowledge. The only time that a channel of communication is working and makes sense is when the transmitter and receiver agree on what the message is. This means that both the transmitter and receiver must agree on a code ahead of time. Then the receiver's problem reduces to figuring out which message was actually sent out of all those messages that might have been sent in the code. Entropy is a measure of his uncertainty in that situation.

Looking at a typewriter keyboard helps to visualize the problem. The keys determine the symbols which can be sent. If a typewritten symbol on a piece of paper is smudged, you can ask a simple question: "Which of the 26 letters of the alphabet has been sent?" Of course, you also have numbers and punctuation marks plus upper and lower case letters to choose from. But the number of symbols is finite and if you are familiar with the keyboard you know in advance what the limited possibilities are. Your only problem is to recognize the symbols. Also, if you recognize the first typewritten symbol in a group of symbols (a word of phrase) you have some clue as to what the next and succeeding letters might be. (If you see a "q," you expect a "u" to follow.)

Shannon showed that there is an optimal coding so you can get the most out of a channel by coding messages properly. He also showed mathematically that there are upper limits in a channel of communications; if you try to exceed those capacity limits, then the channel won't carry anything. It will simply break down.

Let's take some specific analogies to the breakdown of overloaded channels. It can happen easily in the study of a foreign language. If a person talks to you slowly enough in a new language, you can understand him. But if he increases the speed there is an abrupt point at which you can no longer understand anything he says. Your mind loses the ability to sort it out.

Another example is to listen to a tape recording in your own language. As you increase the speed of the tape you find it harder and harder to understand. And then at some abrupt point, you can't understand it any longer. One reason for confusion is the rise in frequency. But basically, the words come too fast for you to receive. The same thing is true when you dictate to a stenographer: she'll stay with you as you increase your speed, but at some abrupt point she'll just stop and look at you because she can't put anything down on paper that makes any sense.

Now let's move the illustrations into decision situations.

On an assembly line, the speed of a man's operations is directly related to the decisions he has to make. If all a man has to do is grasp a red object by its handle and move it to a bench, he can go at a very fast pace, particularly if it's rhythmic.

But suppose the sequence changes so that he's supposed to get not only the red ones, but also the blue ones, if and only if they're accompanied by green ones. Now his speed goes down—he has to make a decision each time. And there's a limit to the speed with which he can take all the information, make a decision, and move.

These ideas also apply to organizations. In any organization, if someone has tried to make it efficient, it will probably be functioning near overload conditions, near the limit of its channel capacities. Parkinson's Law applies here, too: People naturally tend to clog communications channels even if they have to invent messages that aren't really needed.

Consider a decision maker in a real organization—one where the flow of information is far from perfect. We usually assume a man can get more information than he already has on a given problem, and that if he can eliminate some uncertainties, he can make a better decision. But is that information, that greater certainty, worth the cost? Getting better information—say, a detailed report on consumer preferences—might be useful. But the cost of getting that information may exceed the potential increase in profit. It just may not be worth the cost, especially in a highly competitive field where profit margins may be low.

So, for economic reasons, decision makers always have to operate with uncertainties. Rather than complete information, they really need to know what kinds of information to look for. We faced this tradeoff in work on hurricane modification projects. The major question was: How much information do we need to prove we won't make a hurrican worse by tampering with it? I would always like to have the information to show how much better we can make the situation. But the first kind of information is more important to our decisions. It's also less expensive. This example points up the fact that it helps to know what you would do with information before you buy it.

Given uncertainties in decision situations, men need time to grapple with their uncertainties, time to reflect before deciding. If you stack up a number of decision makers—each one stopping to reflect in hopes of eliminating part of his uncertainty—then you have a time delay corresponding to a group of cars that do not start off simultaneously when the light turns green. Entropy, then, is the measure of all those uncertainties and I find it helps to think about how much entropy my decision makers have to contend with.

In general, organizations that foster creativity and dynamism will be high in entropy. Uncertainties will be greater in such an organization; people will not simply be making choices, they will be taking chances. These are the kinds of organizations that most managers of innovation say they want.

All right, then. To think about the long-run development of such organizations we must consider some elementary human factors that go into decision making.

First of all, if we are to use the inherent capabilities of people to be creative, we must allow a range of alternatives. If few alternatives are allowed in a system, then there is no need for creative people—computers and automation can handle systems of limited choice.

Sometimes, of course, we don't want people to be creative; we often deliberately use people in situations of limited choice. When a telephone repairman fixes a line, we want to be sure that the next lineman who comes along doesn't find a nonstandard circuit. In this way, standards and standardizations can be seen as a means of reducing the number of responses people might make. It reduces both creativity and expense.

However, if you want your organization to be creative, then you must give decision centers more alternatives. Which means less control, less certainty for the managers. This certainly applies to research centers where people learn by doing. They must be allowed a great many alternatives, even if the uncertainties bring executive headaches.

Another human element is in the criteria for satisfaction used by the decision maker: His criteria of satisfaction may or may not be the same criteria applicable to the organization. There may easily be conflict between the decision maker's payoff and the long-run payoff for the organization. In this regard, my definition of a bureaucrat is a man whose personal payoff function is inconsistent with the payoff function of the organization—whenever you deal with him, the business of the company, or the government, does not get done. Often his payoff function is continuation of power and position, not satisfaction of others' needs or performance of duties.

Good management tries to make the payoff criteria congruent, but is not always successful. To the extent that these criteria can be made consistent, there will be better decision making. This involves more than having the employee identify with the success of the company. Unless his personal reward structure is the same as the company's reward structure, he may be placed in a terrible position.

For example, if a salesman is offered a bonus for getting a large job completed on time—but he thinks there may be a defect in the equipment—what will he do? One possiblity is to get the job installed on time and earn his bonus; when the defect is discovered, the penalty won't come out of his bonus—somebody else will pay for it with his head. Or should the salesman protect the reputation of the company, report the defect, and lose his bonus?

So when we put people into the picture, we add some basic human complications which, in turn, add immeasurably to the uncertainties with which a decision maker lives.

These ideas also apply to delegation of responsibility. We never know everything about a given situation, but usually we do know something. Sometimes we want to communicate that something to another person so that other person can act in our place as though we were there. How can this best be done?

Staying within Shannon's criteria, we don't want to take up all the channel capacity available. It isn't necessary. If the other man merely says, "Tell me everything," he may have to listen to your whole life story. So a code, a set of agreed upon and limited signals, is needed. Within that code we can communicate efficiently, not telling too much or too little but just enough for the other man to make decisions in our place. But we have to be careful—if the code becomes too refined, we'll have difficulty teaching other men to live and work with it.

These very primitive problems show up in organizations all the time. And they can't be resolved easily; organizations are composed of humans, not electric cables of microwave relays: But the analogies from information theory can be helpful, if not directly transferable.

I think these human factors also mean there is an inherent upper limit to the speed of response of any organization. That limit is determined first by the number of levels at which a decision must be made in a pyramidal structure. It is determined by the number of options people are allowed along the line and whether or not you encourage them to be creative. It also depends on how much uncertainty they have to face and the resources they have for dealing with that uncertainty. Finally, it depends on whether their payoff functions are consistent with those of the organization.

To get around this inherent limit to the speed of response, there are strategies which are both legitimate and dangerous.

Normally, we go through regular channels to get information up or down the line. But in emergencies, the manager has to move down into the organization quickly to locate the key decision maker. This means dealing with him directly.

An executive doesn't dare do this very often. It's equivalent to bypassing levels of management and other decision makers. If it becomes a regular practice, all important decisions are pushed into the front office—and there aren't enough channels to get all the information to make all those decisions.

Decentralization is a good way to cut down channel overload but, of course, response time becomes slower. So it's an art to know when to break the rule about going through channels.

How do I go about breaking the rule? First of all, at an early stage of establishing new staff relationships I tell all the people who report to me that I don't intend to do it regularly. But occasionally I have to go around them quickly for what I consider good reasons. For example, as Assistant Secretary of Commerce, I would sometimes go several echelons down the line to check some points about prospective testimony before Congress. I needed the information quickly. But I told the source to notify people above him that I had contacted him and I started notifying people from the top down.

In that instance things worked out well. But in about half the cases, I have regretted breaking the rule. The emergencies were not usually as critical as I had thought. Furthermore, even I have to learn to live with some uncertainties.

Another strategy in sending signals—a very difficult strategy—is to cut down ambiguity. Here the aim is to make the communication very clear.

At the outset, one should be aware that there are two kinds of ambiguity, either of which can be a block to action. One form of ambiguity is contained in the question, "What are we talking about?" The other form is, "What else do we need to know?" For example, I may be able to describe a roulette wheel to someone who has never seen one; that pins the subject down. But I still can't tell him where the ball will come to rest. Reminder: To eliminate ambiguity, there should be no doubt about what is in doubt.

Ambiguity can creep into any situation and in a strange, almost mystifying way, stymie efforts at problem solving. In such situations, ambiguity often reflects the personal problems of the people involved even though it may seem to focus on factual/informational factors.

I can illustrate this best by recalling an emergency that occurred on a jet engine production line a few years ago. On a day I happened to visit as a consultant, an incredible discovery was made: Three changes had been made in the engine simultaneously. The design and location of the fuel nozzles were changed; the material of the combustion chamber was changed; and the fuel was changed. As a result of one, two, or all three factors, the engines were failing during the first hour of their inspection runs.

A meeting was hastily called to find out from the design engineers whether the fault lay with the changed material, the changed fuel, or the design and location of the fuel nozzles. It quickly became apparent that no one in the conference room was certain enough of the answer to justify an immediate change on the assembly line. The only thing to do was to stop assembly of over a million dollars worth of engines and find out what was wrong and how to fix it.

One engineer proposed to take some samples of the new material and the old material and put them in a test chamber downstream from the flames from both the new fuel and the old fuel. That would show if the difference were due to the changed chemistry of the fuel (a new high-performance fuel known to have given trouble before).

One of the other engineers commented, "I should not be surprised if you fail to get any failures that way." At first it seemed that his statement implied that this was not a good test to run. But did it?

A simple test of a statement is to see what its denial implies. I therefore asked him, "Would you be surprised if we do get failures?" His answer was "No."

At this point it became obvious that his original statement was completely ambiguous. To make this point clear, I said to him: "In other words, you will not be surprised if we do or do not get failures."

If his original statement had any meaning, it was, "I am incapable of being surprised in this matter." Insofar as the engineering problem was concerned, it was irrelevant. But such statements, which in this case almost stopped the first engineer from conducting his tests, can create an interpersonal impasse when action is required. Even though the statement sounded clear and direct, like so many that arise in "technical" meetings it was really ambiguous.

One caution about trying to eliminate ambiguity: It is time-consuming. The message must be carefully composed and generally becomes very wordy. At the receiver's end it will take longer than usual to examine and understand it.

There is another important strategy which requires patience and, for most of us, a new perspective on organizational messages. This strategy is to encourage the expression of uncertainty. Obviously, this goes counter to most of our training in writing, engineering, and management. Yet, the expression of uncertainty is closer to reality that forced pretenses at certainty.

The environment of a racetrack can teach us something about expressions of uncertainty. Handicappers are very clear about their uncertainties; they reduce them to numbers, better known as odds. If you ask a handicapper what he thinks about the next race, he quotes odds which indicate the uncertainties about whether one horse or another will win the race.

In contrast, when a general manager asks his staff about the results of a market study—a form of speculation about the future—he's usually told something like: "In the second year we're going to sell 180,000 gidgets." It sounds like a certainty. If he were told the truth, the uncertain truth, it would sound like this: "Well, in the second year we expect to sell 180,000 gidgets, but that estimate could be off plus or minus 60,000 depending on conditions which are still developing." Now the manager knows more about his reality.

In this regard, young executives face a special dilemma when they try to frame statements about the future. They have two basic problems: 1) They want to convince the boss they know what they're talking about; and 2) at the same time they want to keep him from acting as if that were true. On the one hand, they can get brownie points for being omniscient; on the other hand they can lead the organization into grave trouble by pretending to know more than they really know. Our usual demands for certainty set this kind of organizational trap.

An organization can also be looked at as a formal arrangement for decision making where the decision-makers are separated physically—by walls, streets, or long distances. Organizations get things done by channeling their energy, materials, and information around to the separated physical locations. Thermodynamics gives us a body of knowledge for an objective measure of the flow of energy and materials. (At an extreme, a set of meters can keep touch with all the energy and materials flowing through a refinery. People are almost unnecessary.)

But not so with information flow. It is still not a well-developed science.

In a simple description of information flow, different people at different places in an organization—with different degrees of knowledge—must get similar information. They are then supposed to make decisions or take actions within the alternatives open to them according to policies previously enunciated.

If you have worked out a coherent code and information flow for this design, then organizations can move efficiently. If you don't have a coherent information pattern, then different people in different places in the organization have to stop and think. They have to ask questions. They need clearer signals. If the original message came from on high, from top management, then people may turn to their immediate superiors or to their peers for clarification. And those people may not have any better idea of what was intended in the original message.

Sometimes the problem is seen as "noise in the system." Too much message, too many signals, or contradictory signals lead to ambiguity, blockage, and stoppage. What people normally mean by "noise in the system" is that things are happening in the system which interfere with their receiving. Too many signals are coming over their receiving channels at the same time and they can't make sense out of it all. Noise interferes with their ability to decide and act.

Sometimes this becomes a political situation. A signal-receiver has legitimate political questions: Should I believe this signal? Will the man who gave this signal go away, be replaced, move on—and leave me stuck with his signal? Should I just ignore the whole matter, pretend the signal never reached me, or maybe act as if I don't understand it?

This is how signals in a human system can lead to defensive politics. In specific terms, it's part of the problem in automobile safety. In recent years, the signals have been switched: Designers are now being told that safety features are wanted; they used to be told to cut costs on safety features. One auto company executive I know had a tough time convincing his people that the switch in signals was real. Legitimate defensive questions came up: "He says safety is wanted, but I used to get my rewards and promotions for cutting costs. How can I believe him?" Or, "This man seems strong on safety and antipollution devices, but how long will he have his job? What will the next man want?"

Sometimes a switch in signals evokes a stronger, almost offensive political response. When I took over in my position in the Department of Commerce, one man with years of experience through several administrations told me in advance that he wouldn't pay attention to my new signals. He thought I'd be too busy to pay attention to what he was doing. He also knew that I would not be around forever, that administrators come and go, but civil servants, like Washington correspondents, stay on through many administrations.

Noise in a system then, is a decision problem. Some people get clearer signals from organizational noise than others do because they have enough experience to interpret and predict from it. Furthermore, some people enjoy operating in a state of ambiguity where even if signals are contradictory, they can select the one they want as authority for their decisions or actions.

Still another way I often look at an organization is as an attempt by a group of people to develop a consistency and predictability in their relationships with each other and with outside phenomena. If the system is to have a very high degree of predictability, then it's necessary to narrow down the options for action of the people in it. Their options are usually reduced through policy statements which they are expected to follow. If at the same time an organization is to perform consistently, then it has to be concerned with increasing and/or maintaining its efficiency. There are several ways to consider the efficiency of an organization; not all of them are compatible. For example: How fast can an organization respond? Or given a long period of time, how well can an organization decide what it wants to do? Or how does an organization get the information it needs to judge what it has done?

Very often, efficiency is looked at simply as a function of the first factor; speed of response. In this regard, I've started to think that the speed of response of a system is limited just the way a thermodynamic system is limited in its efficiency—and for the same reasons.

I believe that some day the laws of upper limits to organizational efficiency will be stated in a manner similar to the second law of thermodynamics which covers the upper limits of the efficiency of a heat engine.

The concept of entropy applies to both engines and organizations. We don't have enough information on either steam engines or organizations to do much about their efficiency. With steam engines, we only get information about their pressure and temperature. This is very limited information about the useful work of an engine. But to get more would cost more than it's worth.

If we do the same with a company—restrict our information to such measures of efficiency as a profit-and-loss statement—then there's an upper limit to the efficiency which we can get out of that organization. And the efficiency will be pretty low. As we develop more measures based on information about inputs in an organization—not merely its output performance —then the efficiency may be expected to rise.

For an organization busy generating changes and innovations, we don't really have any satisfactory measures of efficiency. My speculation is that an organization is in good shape when it is doing lots of good things but you can also see lots of things to improve. For example, the 3M Company is overflowing with good ideas that can't all be adopted; it can't get a higher level of efficiency because of limits on space, resources, and people. This kind of apparent creative wastage is probably one sign of a dynamic organization. In thermodynamic systems, peak efficiency and peak power do not occur at the same operating condition. Maybe we ought to ask more about the power (rate of output).

In contrast, in a production situation, where output is the basic concern, you can become more efficient by tightening up. But in a preproduction stage, it's hard for a man to explain or justify all the waste motions in his new creation. If he's pressed, he might hide or stifle his new idea or creation.

I recently found this to be a widespread research characteristic. I asked a Russian visitor from Novosibirsk, "What do you do when you have a good idea, you try to sell it, but you can't get it funded?" Through the interpreter he replied, "I do the same thing you do—cheat. I hide it somewhere until it can fly on its own. Then I bring it forward."

In research, therefore, I think it's important to make your people feel comfortable. It's tough enough for them to be creative—why make them feel a sense of guilt over cheating? I was always ready to sacrifice efficiency for productivity in research, but the Congressional committees did not always share this view—nor did the Bureau of the Budget or most of the government.

Speed of response and efficiency often depend on the range of options available. Clear, limited options bring fast, efficient responses. At least that's true in the short run. Unfortunately, many organizations bankrupt their futures for guaranteed returns in the current fiscal year.

Paul Revere was able to respond quickly because his options were limited: "One if by land and two if by sea." That kind of communication is easy to send and it's easy to control what the man at the receiving end will do. But if Paul Revere had been told, "Look, the redcoats seem to have ten possible strategies that we can identify, and we'd better leave room for some surprises besides," then a more complicated communications code would have been needed at Old North Church. There would have been less certainty that Paul could figure out what all the lanterns meant and it would have taken him longer to get on his horse and ride.

If an organization operates within very narrow limits—only two options —then it can respond very quickly to a new situation. It will work well if the new situation comes with a built-in limit of two options. Then you can program everyone up and down the line to this very narrow range of possibilities, completely limit the flexibility of the organization, but speed up its responses.

But that's not the way most new situations develop and that's not the way most organizations can operate. Events are much more complicated.

If you want an organization that will respond quickly to orders from the top, then you must have either limited signals, or a wider range of signals with careful explanations of what they all mean. This requires a narrow, mutually accepted, and well-established code. But a narrow, well-established code in an organization leads to little or no innovation because the approved responses are already programmed.

In contrast, if you want a creative, innovative organization, then your communications channels have to be broader. You also have to accept the prospect that people have the option of generating alternatives that you didn't think of in your messages. It will be harder for some people to find your signal in the accompanying noise of the broader channels. Others, more experienced or more adventurous, will ignore some signals and respond to .others. As a result, you will be less certain that when you issue an order it will be obeyed as you envision. But those tradeoffs must be made if you want to foster creativity in a large organization.

This is a conflict I faced every day with the 15,000 people in my jurisdiction at the Department of Commerce. At the top, if I sent out a creative idea with new, seemingly random and sometimes contradictory signals, it could disrupt the entire system. Or, at the bottom, if a man got an innovative idea he might push it through successfully only if it didn't call for too many changes too many levels up. But if the idea had to come all the way up to me and go all the way down again, approved at every step, then he had a tough time selling it: He was trying to put a new message into the system and at every step it was noise or overload to somebody. Further, since the signals were new, everybody stopped and talked about them, so it was a slow process.

Even when I want creativity and can learn to tolerate some disruption, there are still communication problems. Looking up from the bottom, a man can legitimately ask: Does Tribus mean what he says? Does he want me to be creative? Or is he full of words and should I stick to the limited options already laid out for me before he arrived?

If I want to convince him that I mean it, then I must assume the responsibility of communicating on broader channels, sending more information with wider options. So I also assume the risks that noise will enter the system. Or that I'll have to work harder to keep channels open.

Here's what I do. I make it a point to go down the line—actually talk to the men "on the bench." Obviously, I can't do this with everything that goes on. But when I think something is important, I want to be briefed at the bottom-most level. This means going four or five levels down into the organization. It also means taking a day or two for a trip to talk to the right man.

I don't settle for a simple briefing. I argue with him about what he's doing for several reasons. First of all it's good human relations. In my effort to understand what's going on there is an implied appreciation of what he's doing. Second, when I return to my office I really know what's going on.

For example, when the Tiros III satellite made the first soundings in space from which we could actually compute the temperature of the atmosphere from the earth's surface on up, I wanted a briefing from the man who invented it. He did a magnificent piece of work. And I learned not only what Tiros III could do, but also what he had to do to make it all possible.

I have a very homey analogy. At times I've made furniture in a home workshop. When visitors came over, the wife would usually say, "Isn't that a nice piece of furniture?" But the man would often turn it over, examine it, and say, "Hey, that's a clever joint. Why did you make it that way?" I knew he appreciated it. Also, he learned from it.

I think it's important to have that kind of relationship with technical people. If an executive doesn't have the knowledge to deal that way with technical people, either he should have someone on his staff who does, or he should learn to do it. He should occasionally become a pupil of the people who work for him.

In such a situation, information flows freely. In contrast, technical people often think, "Why bother giving him the information; he wouldn't understand it." When that happens, they will try to tell you only what it will take to get you to do what they think you ought to do. There will be little information exchanged and you won't understand what's going on.

In my dealings with Congress on budget matters, I faced similar problems of overloading channel capacities with information. In government, you are forced to use a highly unsatisfactory method of talking about the national budget of $200 billion. That huge budget is supposed to be settled in about one month of Congressional hearings. Clearly, it's not possible to fully explain the expenditure of $200 billion in one month of conversations over a table, even with lots of bureaucrats talking in parallel to legislators. $1 million is still serious to me, and there were millions a year going through the organizations in my area.

Because of the limited channel capacities of the receivers of information —the members of Congress who consider the budget—they end up talking only about changes from the previous year.

It's a technique called "incremental budgeting." In this game, when a bureau chief is asked, "How much money do you need for next year?," he can answer, "$8 million more than this year." Now his boss and a Congressional committee can concentrate on the additional $8 million and not expend the time and energy involved in reexamining and rejustifying the basic budget of, say, $40 million. "Incremental budgeting" saves time and channel capacity, but you hardly ever get down to basics on whether or not a program is worth any dollars in the first place.

If a Congressional committee had to go over every item on every page in order to understand every dollar, then there would be channel overload. They would have the kind of breakdown that comes when a foreign language is spoken too rapidly—they would understand nothing. So committee chairmen structure the situations to get only the information they can reasonably receive and act upon.

I've tried to think of alternative ways of conducting budget hearings, as many others have. One way is to review programs only every two years. In this mode, Congressional committees could spend twice as much time on half as many items. Another way is to selectively go to "zero-based" budgeting; that is, selectively review entire programs about every five years.

Naturally, these would upset established procedures in the Congress. But I do believe we need a new method of budgeting. Of course, any new method must take account of the cost of listening time and the limit on available channels.

To conclude my speculations about information and decisions in organizations, I want to consider a difficult, but vital, question: How do you tell someone something he doesn't want to hear? This is an unresolved, perhaps unresolvable, problem in information theory.

All executives face this problem. Like kings of old, they could kill (demote, fire) all messengers who bring them unwanted news. But this policy would not help them to deal with reality and it would guarantee an end to the information needed to deal with reality. I really have never solved this problem for myself, as communicator, but I hope I make it easier on those who report to me.

If people tell you something you don't want to hear—don't ever let them know you didn't want to hear it. Otherwise they won't tell you any more. The worse the news, the more concern you should pay to the feelings of the news-bearer. You thus may avoid a limited choice of hearing nothing, or hearing only things you want to hear. Sometimes, as we all know, bad news is important.

I include uncertainty in this category. If your employees really don't know all the answers, they have to be able to admit it so you can get the benefit of what they do know. And this has to be done—can only be done—without the fear their heads will roll because they don't have all the answers.

Entropy is in the decision-maker's head. If you want an organization in which there is no entropy—no uncertainty, no choice—you can create one. You can clear your head. But in my view, you will also clear your organization of the people who can keep it creative and dynamic. Without entropy, you will have a dead system.

September 1970

Index